COMPUTATIONAL FLUID DYNAMICS WITH MOVING BOUNDARIES

WEI SHYY
Department of Aerospace Engineering
University of Michigan

H. S. UDAYKUMAR
Department of Mechanical and Industrial Engineering
The University of Iowa

MADHUKAR M. RAO
ACRi Infotech Pvt. Ltd.
Bangalore, India

RICHARD W. SMITH
Research Engineer
Naval Surface Warfare Center
Panama City, Florida

DOVER PUBLICATIONS, INC.
Mineola, New York

Bibliographical Note

This Dover edition, first published in 2007, is a slightly corrected, unabridged republication of the work originally published in the Taylor & Francis Series in Computational and Physical Processes in Mechanics and Thermal Sciences by Taylor & Francis, Philadelphia, Pa., in 1996.

Library of Congress Cataloging-in-Publication Data

Computational fluid dynamics with moving boundaries / Wei Shyy . . . [et al.].
    p. cm.
    Originally published: Philadelphia : Taylor & Francis, c1996, in series: Series in computational and physical processes in mechanics and thermal sciences.
    Includes bibliographical references and index.
    ISBN-13: 978-0-486-45890-8 (pbk.)
    ISBN-10: 0-486-45890-3 (pbk.)
    1. Fluid dynamics—Technique. 2. Computer algorithms. 3. Numerical grid generation (Numerical analysis) I. Shyy, W. (Wei)

QC151.C66 2007
620.1'064—dc22

2006048825

www.doverpublications.com

To Our Families

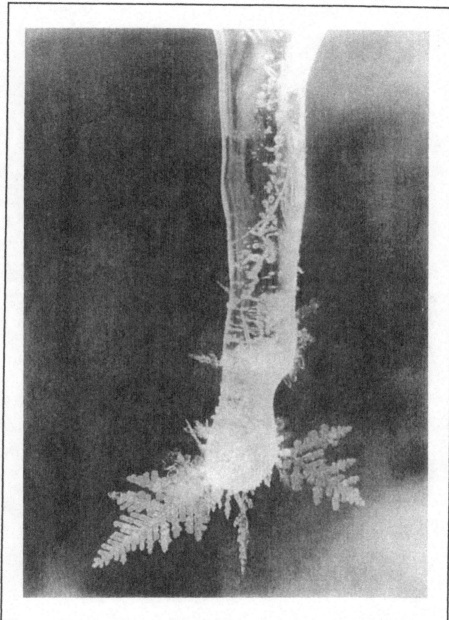

Frost crystals on icicle, Otesgo County, New York. Photograph by
Stephen P. Parker (1954–1990). Printed with permission by
Ms. Yana Parker.

CONTENTS

PREFACE

Many physical phenomena of interest to the fluid dynamics community must contend with the presence of a moving boundary. Since the interaction of fluid flows with moving boundaries leads to a highly coupled, nonlinear system, such problems have remained analytically intractable. The main difficulty is that the moving boundary itself must be determined as part of the solution of the system of equations which govern the behavior of the fluid flow field. In spite of the complexity of the problem, modern analytical, computational, and experimental techniques have substantially enhanced our understanding of a significant number of moving boundary problems. Currently, several books treating mathematical and engineering aspects of moving boundary problems are available, including, for example, Crank (1984), Rubinstein (1971), Pelce (1988), Slattery (1990), and Zerroukat and Chatwin (1994); however, none of them discusses the computational aspects in the context of solving the complete set of the governing equations of fluid dynamics.

The goal of writing this treatise is to present recent developments in computational techniques pertaining to moving boundary problems in fluid dynamics. It continues the discussion presented in a previous book by Shyy (1994) which focuses on interfacial transport. The emphasis here is on the development of computational algorithms and their application to practical engineering problems in a broader context. Furthermore, our intent is to demonstrate the application of a variety of techniques for the numerical solution of moving boundary problems within the framework of the finite-volume approach. In order to illustrate the computational and theoretical issues involved, examples arising from high-temperature materials processing and fluid-flexible structure interaction are chosen for detailed coverage. There are, of course, many more physical phenomena involving moving boundaries, and numerical techniques for treating them, to which we cannot do justice within the scope of this book. In Chapter 1, we summarize the major computational techniques reported in the literature along with a brief account of some physical applications. The methods reviewed in Chapter 1, although not presented in detail, complement those developed in the book.

In order to enable the reader to follow the detailed numerical development, the pressure-based algorithm for fluid flow equations is summarized in Chapter 2. This chapter mainly serves to highlight some of the important computational aspects that have not been covered in detail in many texts on computational fluid dynamics. However, most of computational techniques for handling moving boundaries presented in this book can be directly employed in other fluid flow solution algorithms, and are not restricted to any particular flow solver. In the context of each topic discussed, we have made an effort to systematically account for the entire scope of the computational task, including formulation, nondimensionalization and scaling, algorithm implementation, and physical interpretation. For example, in the area of materials processing, interaction among convection, conduction, and capillarity has been stressed. Both free and internal boundaries are addressed within the context of fixed grid, moving grid, and combined Eulerian-Lagrangian techniques. Furthermore, modeling issues associated with scale disparities arising from fundamental physical mechanisms are also presented. For fluid-flexible structure interaction problems, high Reynolds number but still laminar flow past a flexible, elastic membrane wing is investigated. Moving grid techniques in conjunction with curvilinear coordinates are employed here.

No effort has been made in this book to address any of the generic aspects of computational fluid dynamics; numerous books dealing with these aspects have already been published; they are cited in the text. Instead, we concentrate on those aspects which are primarily related to moving boundaries, and present the details of implementation for the numerical algorithms. We hope to impress upon the reader that it is imperative to recognize the level of complexity associated with each interface tracking approach and to make an appropriate choice based on the physics of each particular problem. It is also hoped that this work can aid other researchers interested in simulating moving boundary problems to appreciate the relative merits of these methods and, if necessary, to be able to develop actual computational tools to perform the needed tasks.

We are grateful for the contributions made by our collaborators here at the University of Florida. In particular, some of the research conducted by Ms. Hong Ouyang has been included; Dr. Siddharth Thakur, Dr. Jeff Wright, Dr. S.-J. Liang, and Professor Craig Saltiel read our manuscript and offered valuable suggestions in numerous places. The work reported has been supported during the last several years by General Electric Company, the National Science

Foundation, NASA Space Grant, the Air Force Office of Scientific Research, and BDM Federal. To these organizations we express our gratitude. We also acknowledge the helpful discussions held with our colleagues, Dr. Dan Backman and Dr. Dan Wei of GE Aircraft Engines, Dr. Yuan Pang of BDM, and Professor Roger Tran-Son-Tay of the University of Florida.

Notes added before second printing

We appreciate the encouragement and warm response received from many colleagues after this book was published. Typographical errors pointed out by readers and spotted by ourselves have been corrected.

February 1997

Notes added before third printing

We have corrected several more typographical errors. Needless to say, we continue to welcome your comments and suggestions.

July 1998

NUMERICAL TECHNIQUES FOR FLUID FLOWS WITH MOVING BOUNDARIES

1.1 INTRODUCTION

1.1.1 Motivation

Moving boundary problems arise in a variety of important engineering applications (Crank 1984, Zerroukat and Chatwin 1994). Typical examples are materials processing, fluid-structure interactions, flame propagation and oil recovery. These systems are characterized by internal boundaries or interfaces demarcating regions with different physico-chemical properties. Across these interfaces, compositions, phases, material properties and flow features can vary rapidly. The interfaces move under the influence of the flowfield and in turn affect the behavior of the flow. Under certain conditions, usually characterized by one or more control parameters, the interfaces may experience instabilities. Such an event–typically a cascade of interfacial instabilities on an initially planar or smooth front–often leads to the formation of patterns and morphological structures, due to the interaction of the competing physical mechanisms present in the system. The prediction of the dynamics of such unstable, evolving interfaces is important in many technological applications. For example, the formation of deep cells in the solidification processing of multi-component materials is quite important since the resulting compositional inhomogeneities will affect the performance and properties of doped semiconductors and the structural integrity of alloy materials (Ostrach 1983, Langlois 1985, Brown 1988, Shyy 1994). Another example is the interaction of fluid flows with non-rigid structures, where fluid convection and deformation of the structure strongly interact, resulting in a highly nonlinear system (Marchaj 1979). Contact melting (Sparrow and Myrum 1985, Moallemi et al. 1986, Hirata et al. 1991) is another example; as recently reviewed by Bejan (1994), this topic has received much attention in the area of tribology (Pinkus 1990).

In moving boundary problems, not only are the transport of momentum, heat, and mass coupled, but also the formation, evolution, and dynamics of the interface play major roles in defining the behavior of the system. Apart from the inherent nonlinearity of these diverse phenomena, the interfacial deformation in itself can be a highly complicated and analytically intractable feature. Several physical phenomena involve interfaces that, as a result of instabilities or external impulses, undergo severe deformations, such as dilatation, compression, fragmentation, and collision. The physics involved in such processes impose severe demands on the numerical, analytical, or experimental approaches used in their study.

For example, during crystallization the interface separating the solid from its melt or vapor can undergo morphological changes that can have serious technological implications (Tiller 1991a) as well as aesthetic appeal (Bentley and Humphreys 1962). These changes of morphology can contribute to the inhomogeneous distribution of the solute that is rejected from the solid into the melt. The rejected solute accumulates in the regions around the irregularly shaped phase front and both microscopic and macroscopic irregularities in the solute distribution result. This phenomenon of solute rejection and transport is called micro– or macro–segregation depending on the scale at which such processes occur. It is evident that in producing crystals and alloys for technological applications, such behavior of the solute field is undesirable and thus has to be minimized or eliminated. The understanding and observation of these phenomena are however hampered due to the extremely disparate length scales involved (ranging from a few microns to meters) and the inherent nonlinearity of the problem rendering analysis very difficult. Instabilities during solidification lead to highly deformed geometries, such as cellular and dendritic forms. In the events leading to complete solidification, a series of topological changes occur.

The interconnected solid matrix encloses isolated liquid pools, which solidify upon cooling. In the case of dendritic branch detachment (Sato et al. 1987, Heinrich et al. 1991, Neilson and Incropera 1991), or in equiaxed or globulitic crystal forms (Beckermann et al. 1994, Ni and Beckermann 1993), convection and settling of isolated crystallites significantly influences the final grain structure. In some experiments, periodic wavelength readjustment of cellular phase fronts has been induced by changing the growth conditions (Cladis et al. 1990, de Cheveigne et al., 1990, Flesselles et al. 1991). Macrosegregation resulting from formation of freckles (Felicelli et al. 1991, Flemings 1974), of channels (Neilson and Incropera 1993) and positive and negative segregated bands (Diao and Tsai 1993) can significantly impact on the quality of the alloys. Shrinkage-induced convection can also have noticeable effects on the solidification process (Chiang and Tsai 1992a, b). It is clear that in order to follow the entire spectrum of possible phenomena in the solidification process, it is necessary to cope with a variety of topological changes. A recent review by Prescott and Incropera (1993) has summarized from an engineering point of view the interaction between fluid flows and macrosegregation. Beckermann and Viskanta (1993), Langlois (1985), Samanta et al. (1987), Shyy (1994), and Viskanta (199) have given accounts of mathematical and computational modeling of transport processes during solidification of alloys.

In liquid–gas systems, the formation of sprays (Bogy 1979, Lefebvre 1989), the shattering of droplets, and other surface instabilities (Fromm 1981, Orme and Muntz 1990, Spangler et al. 1995) present challenges in tracking multiple, interacting interfaces. In practical systems such as propulsion devices (Lefebvre 1989) and internal combustion engines (Kuo 1991, Lai and Przekwas 1994), since the phase change process takes place in chemically reacting flows (Rosner 1986, Williams 1985), these problems become extremely difficult to solve. In each of these phenomena, the necessary boundary conditions at the interface are functions of the interface shape, since the solution in the entire domain critically depends on the accurate determination of the interface shape and its derivatives.

Typically, the combination of interfacial dynamics and physico-chemical transport processes is at the heart of moving boundary problems. Due to this interaction, as well as to the presence of other complicating mechanisms mentioned above, moving boundary problems often are very difficult to analyze. The difficulty exists even though the governing laws describing interface characteristics are well established from the fluid dynamic and thermodynamic viewpoints. The main difficulty is that the internal boundary position and shape must be determined as part of the solution of the transport equations of mass continuity, momentum, and energy. Within each domain, the field equations must be solved with the location of the internal boundary being determined simultaneously. Under certain conditions the interface may undergo a succession of deformations leading to very complicated shapes. Finally, from the continuum mechanics viewpoint, the interface often is treated as a discontinuity in the flowfield. Within the inevitable limitation of finite grid resolution, this discontinuity needs to be accurately tracked both in time and in space.

1.1.2 Overview of the Present Work

In this work, several issues related to devising solution techniques for moving boundary problems are presented. Therefore, it is appropriate first to provide a brief survey of some existing techniques concerned with tracking highly distorted fronts, including both Lagrangian and Eulerian methods. A variety of methods is available, and as will be evident from the applications detailed later in this work, the choice of an efficient and robust method will depend on the physical problem under investigation.

To illustrate the computational and theoretical issues involved, examples arising from high temperature materials processing and fluid-structure interaction are chosen to give detailed coverage. Of course, there are many more physical examples involving moving boundaries to which we can not do justice within the scope of this book. In the context of each physical topic discussed, we have made an effort to systematically account for the entire scope of the computational task, including formulation, nondimensionalization and scaling, algorithm implementation, and physical interpretation. In particular, besides the development of numerical algorithm, other aspects also will be discussed to help highlight the issues encountered. For example, multiple scaling is a well-known feature in many mechanics problems (Ciarlet and Sanchez-Palencia 1987); it is also a prominent feature of moving boundary problems. This aspect will be addressed theoretically as well as computationally.

In order to supply necessary information to enable the reader to follow the detailed numerical development, the pressure-based algorithm for fluid flow equations is summarized in Chapter 2. This chapter serves mainly to highlight some of the important computational aspects that have not been covered in detail in many texts on computational fluid dynamics. However, most of the computational techniques for handling moving boundaries presented in this book can be directly employed in other fluid flow solution algorithms and are not restricted to any particular fluid flow equation solver employed here. The formulation presented in Chapter 2 is applicable to generalized curvilinear coordinates, with a moving, adaptive grid for the discretization of the physical domain. The Cartesian, fixed grid formulation then reduces to a special case of this more general procedure. To prepare for the later chapters, formulation and computations of Marangoni convection (Chen 1987, Koschmieder 1993, Levich 1962, Probstein 1989, Shyy 1994) also are discussed in Chapter 2. Marangoni convection is associated with flows that arise due to interfacial tension gradients and density gradients. These flows play a pivotal role in the processing of single crystals from a melt and therefore significantly influence crystal quality. The interaction of steady-state Marangoni convection and the material properties is rich in structure and pattern.

In Chapters 3 and 4, a moving grid technique is presented along with a curvilinear grid-based pressure correction algorithm. In this approach, the interface is explicitly tracked at every instant, with the individual domains mapped by curvilinear grids. Grid remeshing is conducted at each time step to conform to the moving boundary. Within each domain, the pressure-based algorithm is utilized to solve the fluid flow equations, in combination with the boundary conditions pertaining to interfacial behavior. To illustrate the performance of this approach, two physical problems are presented, namely a phase change problem at the morphological scale involving the balance between capillarity and heat conduction, and the interaction between a viscous flow and a flexible membrane wing. For problems involving interaction of fluids and structures, high Reynolds number but still laminar flows surrounding a flexible membrane wing are investigated. This grid mapping technique carries with it the advantage of using the explicit interface information in applying the boundary conditions in each phase. However, it encounters difficulties when the interface becomes multiple valued or geometrically complicated. For a curvilinear grid, this presents difficulties in conforming to the interface, since a highly contorted boundary can lead to excessive grid skewness. When the interface exhibits topological changes such as breakups and mergers, a mapping technique faces fundamental difficulties. Massive grid restructuring and reordering may be called for, leading to an insurmountable algorithmic burden.

In Chapters 5, 6 and 7, we address fixed grid techniques. First, we discuss an Eulerian method, known in the current heat transfer literature as the enthalpy formulation. Specifically, the content of the physical problem discussed in Chapter 4 is broadened here; interaction among convection, conduction, and capillarity is stressed. Both free and internal boundaries are addressed. Furthermore, modeling issues associated with scale disparities arising from geometrical complexity and fundamental physical mechanisms also are presented. In Chapter 5, the interface is handled implicitly with the two-phase region modeled as a porous medium. The liquid fraction varies smoothly across this porous, so-called "mushy region". In such a formulation, the interface occupies a finite region instead of being defined as a discontinuity. This allows a unified set of governing equations to represent the transport processes in the whole domain. The mushy zone, or interface region, is modeled via the phase fractions, which are incorporated into the source terms in the governing equations to account for the phase change phenomena. Obviously, both the unified mathematical structure and the fixed grid computational procedures make this approach attractive. However, lack of precise definition and details of the interface makes this approach unsuitable for certain classes of problems. For example, when the capillary effect becomes important, estimation of the interface curvature is of paramount importance, as dictated by the Gibbs-Thomson effect. Under such circumstances the fixed grid approach is not the method of choice.

To combine the strengths of the moving grid and fixed grid techniques, advances have been made recently in the area of combined Lagrangian-Eulerian methods, which are described in Chapters 6 and 7. In this approach, a set of markers is employed to define and follow the interface in a Lagrangian framework. To facilitate the solutions of the field equations, a fixed grid is utilized. On the fixed grid system, the markers advance in time, causing the computational cells in the interface regions to become irregularly shaped. Special treatment is needed to enable accurate computations of the mass, momentum, and energy fluxes and to cast the discretized forms within a pressure-based, control volume framework. In the context of solidification problems, both the morphological evolution subject to the Gibbs-Thomson effect (i.e., when capillarity and heat conduction are dominant), and the macroscopic interface dynamics under the influence of convection and conduction are discussed. Also presented are direct comparisons between the fixed grid, cut-cell approach and the curvilinear grid algorithm for solving highly convective fluid flow problems in an irregularly shaped domain.

The techniques presented here can be used for a wide variety of fluid flow problems with moving boundaries. Many aspects associated with these techniques have yet to be addressed to their full extent; it is our hope that this book can help stimulate progress in this exciting area.

1.2 NUMERICAL METHODS APPLIED TO GENERAL MOVING BOUNDARY PROBLEMS

Several techniques exist for tracking arbitrarily shaped interfaces, each with its own strengths and weaknesses (Crank 1984, Floryan and Rasmussen 1989, Shamsundar and Rooz 1988, Wang and Lee 1989). These techniques may be classified under two main categories: (a) surface tracking or predominantly Lagrangian methods (Harlow and Welch 1965, Vicelli 1969, Chen et al. 1995) and (b) volume tracking or Eulerian methods (Hirt and Nichols 1981, Ashgriz and Poo 1991). The main features of the two types are presented in Fig. 1.1.

In the Lagrangian methods, the grid is configured to conform to the shape of the interface, and thus it adapts continually to it. The Eulerian methods usually employ a fixed grid formulation, and the interface between the two phases is not explicitly tracked but is reconstructed from the properties of appropriate field variables, such as fluid fractions. Based on these basic differences in approach of the two classes of methods, the following comparisons can be made:

1. Interface Definition

The Lagrangian methods maintain the interface as a discontinuity and explicitly track its evolution. If detailed information regarding the interface location is desired, Eulerian methods may need elaborate procedures to deduce the interface location based on the volume fraction information, and uncertainty corresponding to one grid cell is unavoidable (Ashgriz and Poo 1991, Hirt and Nichols 1981, Lafaurie et al. 1994, Liang 1991). In the Lagrangian case, the interface can be tracked as an $(n-1)$-dimensional entity for an n-dimensional space (De Gregoria

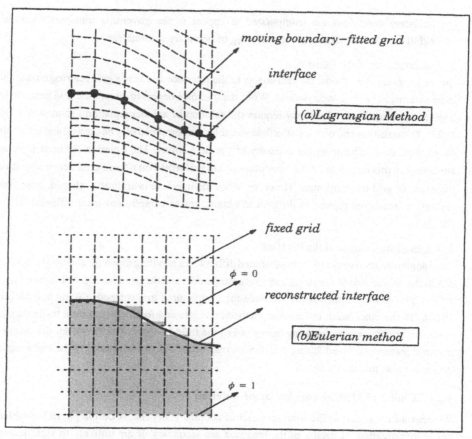

Figure 1.1. Comparison of Lagrangian and Eulerian methods for interface tracking. (a) Purely Lagrangian method with a moving, boundary conforming grid. (b) Fixed grid Eulerian method with a phase fraction definition of the interface.

and Schwartz 1985, Glimm et al. 1988, 1986, Miyata 1986, Wand and McLay 1986). No modeling is necessary to define the interface or its effect on the flow field. In the case of Eulerian schemes, modelling or the solution of additional equations is required to obtain information regarding fluid fractions or other functions yielding information in the two-phase regions.

2. Interfacial Boundary Conditions

In the Lagrangian methods, boundary conditions can be applied at the exact location of the interface since the interface position is explicitly known at each instant. In the Eulerian methods, the boundary conditions are manipulated to appear in the governing transport equations (Brackbill et al. 1992). This leads to smearing of boundary information.

3. Discretization of the Domain

In the Lagrangian methods, the grid adapts to the interface and hence grid rearrangement and motion terms have to be incorporated. When the interface begins to distort, the grid needs to be regenerated each time, which may require the solution of another equation (Thompson et al. 1985). The resulting grid on which the field variables are computed may be skewed and unevenly distributed, thus influencing the accuracy of the field solver. The Eulerian methods have an advantage in this regard since the computations are performed on a fixed grid, hence obviating the need for grid rearrangement. However, when the interface is arbitrarily shaped, improved resolution in desired regions is difficult to obtain, unless complicated local refinements are adopted.

4. Topological Changes of the Interface

Lagrangian methods have so far experienced difficulty in handling topological changes, mainly due to the breakdown of the structured grid arrangement and the need for redistribution of field information in the vicinity of the interface for unstructured grid methods (Wang and McLay 1986). On the other hand, in Eulerian methods mergers and fragmentations are taken care of automatically, merely by updating the values of the fluid fraction field. However, the detailed physical features involved during such events may not be fully resolved due to the smearing of information as mentioned above.

1.2.1 Choice of Method-Lagrangian or Eulerian ?

Whether a Lagrangian or Eulerian approach is adopted is dependent on the physical problem under investigation. If details of the interface are secondary or are unlikely to significantly impact the global flow features, the Eulerian methods are more attractive. If the discontinuity across the interface is to be maintained with fidelity, and if interfacial behavior is the focus, Lagrangian methods hold an advantage. Hitherto, Lagrangian methods have been impeded by difficulties associated with topological changes in the interface. If it were possible to combine the strengths of the fixed grid with explicit tracking of the interface and devise a scheme to circumvent the logical hurdles in the events of interface mergers or breakups, a method of broader applicability would result.

1.2.2 Review of Available Methods for Moving Boundary Problems

Under the broad categories of Lagrangian and Eulerian methods, the following numerical techniques have been developed thus far by researchers in the area of moving boundary problems.

1.2.2.1 Transformation Methods with Body-Fitted Coordinates

In these methods the irregular physical boundary is mapped by body–fitted, but structured, meshes on which the field equations are solved and moving boundaries tracked. In Chapter 3, we will devise such a body-fitted moving grid solver using generalized curvilinear coordinates and present case studies, namely a flexible membrane wing in unsteady flow and a morphologically unstable solidifying front. However, as with most mapping methods, the calculations experience difficulties when the interface becomes multiple-valued (Brush and Sekerka 1989). It is still possible to generate boundary-conforming grids beyond this stage, say by solving partial differential equations in each phase (Thompson et al., 1985); however, the added expense of solving these equations is undesirable. Several techniques have been developed to track the moving boundaries, including the Lagrangian–Eulerian formulation (Hirt et al. 1974), deformed–cell method (Mizuta 1991), and marker-and-cell method (Chan and Street 1970, Daly 1969, Vicelli 1971).

Boundary-fitted grids often experience difficulties in the form of grid skewness under severe interface convolution and need to be reconfigured under topological changes of the interface. Such events need to be identified and dealt with–a process that will involve considerable logical and algorithmic complexity. Furthermore, the grid points and values of the field variables have to be redistributed in the vicinity of the interface, which may lead to additional numerical dissipation.

1.2.2.2 Boundary Element Methods (BEM)

Due to the facility offered by the BEM in terms of the reduction of the dimensions of the problem (Brebbia et al. 1984, Chen and Zhou 1992), it has been extensively employed in front tracking. Dendritic structures (Saito et al. 1988, Strain 1989) and Saffman-Taylor fingers (DeGregoria and Schwartz 1985, 1986) have been simulated with considerable success. The BEM is well suited for problems with linear field equations, but its application to nonlinear problems with convection and for physical situations involving time-dependent boundary conditions (Bouissou et al. 1990, Fabietti et al. 1990) and topological changes is still under intensive development. Recently, Spangler et al. (1995) have applied the BEM to simulate nonlinear instabilities during jet atomization processes in the wind–induced regime; in this work, the unsteady liquid and jet behavior is assumed to be governed by the Laplace equation, i.e., the fluid flow is considered to be incompressible and inviscid, so that the viscous effect is not considered.

1.2.2.3 Volume Tracking Methods

Volume–tracking (Eulerian) methods differ fundamentally from surface-tracking (Eulerian-Lagrangian) methods discussed above in that the interface is not explicitly defined or tracked but is reconstructed at every step (Hirt and Nichols 1981, Smith 1981, Youngs 1984). The Volume-of-Fluid (VOF) method makes use of a fluid fraction variable f, assigned values of 1 and 0 in the two phases, which is calculated as a field variable over the domain. The interfacial cells are then identified as those with fractional values of f. The volume fraction is then advected with the local flow velocity. Over the years, more accurate schemes for advecting the volume fractions have been developed extensively for dealing with passive liquid-gas interfaces · (Brackbill et al. 1992, Ashgriz and Poo 1991, Liang 1991, Kothe and Mjolsness 1992, Youngs 1982). As reviewed by Kothe and Rider (1994), in general, two classes of algorithms have been employed, namely, piecewise constant and piecewise linear, yielding different degrees of accuracy, to recover the interface shape. There also has been effort of combining the VOF method with the finite-element method to simulate solidification and filling processes (Zhang et al. 1994).

The most striking disadvantage of the volume tracking schemes is that while mergers and breakups of the interface can be handled, they cannot be treated with precision. The main difficulty arises in the reconstruction of the interface which involves a considerable number of logical operations. Only recently has a technique to impose surface tension effects in an efficient manner been developed (Brackbill et al. 1992) in which the boundary conditions on the interface are assigned in a weighted fashion to the computational nodes in the vicinity of the interface, instead of being applied directly on the boundary. Volume-tracking methods have been applied to complex interfacial phenomena, including droplet dynamics and breakup, morphological instabilities in crystals (Smith 1981) and spray dynamics (Liang 1991). In applications where only the broad features of interfacial behavior are required the VOF methods have been widely used. The ability of this approach to accurately resolve an irregularly shaped interface needs to be improved. Recent work by Richards et al. (1994), combining the VOF method and the Continuous Surface Force (CSF) algorithm (Brackbill et al. 1992, Kothe and Mjolsness 1992) to simulate the breakup of liquid-liquid jets has advanced the current state of the art.

1.2.2.4 The Level-Set Method

Notable among new approaches to the solution of interface evolution problems (in particular to morphological instability problems) is the level-set Hamilton-Jacobi formulation (Osher and Sethian 1988, Sethian 1990, Sethian and Strain 1992). We now describe the basic features of this algorithm.

As shown in Fig. 1.2, a variable called the distance function ϕ is defined on the domain, where initially $\phi(x,t)$ = distance from x to $\Gamma(t=0)$, the initial interface location. On this level set,

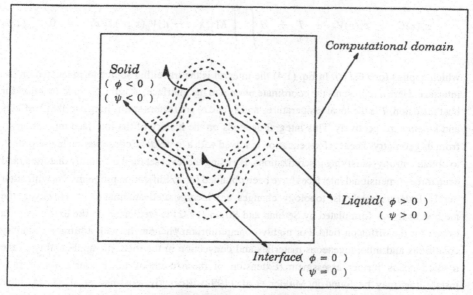

Figure 1.2. Illustration of computational domain and definition of distance function f for the level-set formulation and phase variable y for the phase field method.

the interface at any instant corresponds to the contour $\phi = 0$. A particle on the level set is advected according to the equation

$$\frac{\partial \phi}{\partial t} + \frac{\partial \vec{x}}{\partial t} \cdot \nabla \phi = 0 \qquad (1.1)$$

and

$$\frac{\partial \vec{x}}{\partial t} \cdot \vec{n} = F \qquad (1.2)$$

where F is a speed function evaluated at each point on the Cartesian grid. Thus, the equation governing contour evolution is

$$\frac{\partial \phi}{\partial t} + F|\nabla \phi| = 0 \qquad (1.3)$$

and the initial interface shape as defined by the contour $\phi(x, t=0) = 0$ is specified. In the work of Sethian and Strain (1992), F, the speed function, is a globally defined variable, which reduces to V_n, the normal speed of the phase front contour for $x \in \Gamma(t)$. F is obtained by extending smoothly the expression for $V_n(x, t)$,

$$\varepsilon_c(n)C \ + \ \varepsilon_v(n)V_n \ + \ T \ + \ H\int\limits_t \int\limits_{\Gamma(t')} K(x,x',t-t')V_n(x',t')dx'dt' \ = \ 0 \qquad (1.4)$$

which applies for $x \in \Gamma(t)$. In Eq. (1.4) the integral is the travelling source term located on the interface. Here n represents the coordinate normal to the interface, K is the Greens' function for heat diffusion, T is the local temperature, and ε_c and ε_v are factors controlling capillarity effects and kinetics, respectively. This integral depends on the history of the interface motion, apart from its geometry. The level-set equation is solved with a hyperbolic conservation law approach to obtain entropy-satisfying discretizations. By using the level set method, highly distorted, and even three-dimensional interfaces have been obtained for solidification problems. As with other purely Eulerian methods, topology changes are incorporated automatically. However, the method as it was formulated by Sethian and Strain (1992) is predicated on the existence of a kernel for the diffusion field. For highly nonequilibrium phenomena with arbitrary boundary conditions and inhomogeneous materials and convection in the melt, the applicability of the method needs further assessment. Extension of these methods to another class of flow instabilities may be found in Mulder et al. (1992). Recently, Sussman et al. (1994) have presented results for incompressible two-phase flow based on the level-set approach. Their solutions appear to exhibit high fidelity. However, in this work, while the interface remains sharp without ever having to explicitly find the front, the interface is still of a finite thickness which is artificially assigned. This aspect of the interface treatment is similar to that conducted by Unverdi and Tryggvason (1992), as will be briefly reviewed next.

1.2.2.5 Moving Unstructured Boundary Conforming Grid Methods

In a recent effort, Unverdi and Tryggvason (1992) have employed a moving unstructured boundary-conforming grid in combination with a stationary Cartesian grid to track the motion of bubbles undergoing severe deformation. The method is similar in spirit to the front tracking schemes applied by Glimm et al. (1986) to the Rayleigh-Taylor instability. In each case, the boundary-conforming requirement of the grid leads to grid readjustment procedures close to the interface. In addition, in the study of Unverdi and Tryggvason (1992), following Fauci and Peskin (1988), the surface-tension effects on the interface are transmitted in the form of a force f_s to the underlying Cartesian grid. The grid redistribution then is done based on a weighting function. Thus, although these boundary-conforming grid methods avoid numerical dissipation at the interface, with respect to the field solver, the interface location still is not exactly specified.

Besides the above-mentioned techniques, finite element methods have been developed by several research groups for free and moving boundary problems, e.g., (Lynch 1982, Soulaimani et al. 1991) employing different techniques, such as the combined

Eulerian-Lagrangian formulation (Huerta and Liu 1988, Nomura 1981, Ramaswamy and Kawahara 1987), and the deformable-spatial-domain/space-time procedure (Hughes and Hulbert 1988, Nguyen and Reynen 1984, Tezduyar et al. 1992a,b). In the combined Eulerian-Lagrangian method, mesh quality control is the central issue for problems with large deformation. In deformable-spatial-domain/space-time procedure, the governing equations are written in the space-time domain; therefore, the deformation of the spatial boundary can be automatically accounted for. It has been demonstrated that minimal numerical dissipation is generated, provided that numerical stability can be controlled.

In the area of physical applications, very complex physical problems have been simulated using these methods. For example, Nobari and Tryggvason (1994a,b) have extended the method reported by Unverdi and Tryggvason (1992) and simulated the drop collision and coalescence phenomena. Fukai et al. (1993, 1995) have also developed and applied a finite element method to model the droplet spreading and colliding process. Very interesting results have been reported by these researchers. From a different angle, Voller and Peng (1994) recently proposed a deforming finite element method in conjunction with the enthalpy formulation (Crank 1984, Zerroukat and Chatwin 1994, and to be discussed in Chapter 5), to handle the phase change problem. In their approach, an iterative solution procedure consisting of (i) a fixed space grid scheme that tracks the phase front based on the enthalpy formulation, (ii) a front tracking scheme that constraints the phase front to lie along a chosen grid line on a continuously deforming finite-element space grid, and (iii) an approximate front tracking scheme that constrains the phase front to lie within specified control volumes, is developed. They have applied this method to some simple cases to demonstrate its performance. Independently, de Groh and Yao (1994) have combined the enthalpy formulation and the finite element method to simulate the solidification process of succinonitrile in both two- and three-dimensional configurations.

In the area of finite element methods, domain decomposition has also been applied by e.g., Kassemi and Naraghi (1994), Lynch (1982), Ungar and Brown (1984)to capture the phase boundary; in the study of Ungar and Brown (1984) the deep cell regime has been investigated. The restrictions mentioned above regarding the need for remeshing also apply to these simulations. The finite element method has also been applied to solve practical materials processing problems such as turbulent fluid flow and heat transfer in continuous casting (Thomas and Najjar 1991). Overall, the area of unstructured grid techniques has been undergoing rapid progress with efforts made by many researchers.

1.2.2.6 Phase Field Models

The phase field method (Kobayashi 1993, Wheeler et al. 1993) recently has received much attention and has succeeded in generating realistic solidification microstructures. In this method, as shown in Fig. 1.2, a phase field variable ψ is defined on a fixed grid. The variable ψ assumes discrete values of 1 in the solid and 0 in the liquid. The distribution of the variable is as shown in Fig. 1.3.

The basis of the method lies in expressing the free energy of the system as a Cahn-Hilliard functional (Cahn and Hilliard 1958, Penrose and Fife 1990). For a general non-uniform system it is expected that the local free energy f per molecule will depend both on local composition and on the composition of the surrounding environment. Thus, f can be expressed as the sum of the local composition c and the local composition derivatives, respectively. If f is a continuous function with respect to c and the gradients of c, then a Taylor series expansion around f_o, the free energy of a solution of uniform composition, yields

$$f(c, \nabla c, \nabla^2 c, ...) = f_o(c) + \sum_i \left[\frac{\partial f}{\partial(\frac{\partial c}{\partial x_i})}\right]_o \left(\frac{\partial c}{\partial x_i}\right) + \sum_{ij} \left[\frac{\partial f}{\partial(\frac{\partial^2 c}{\partial x_i \partial x_j})}\right]_o \left(\frac{\partial^2 c}{\partial x_i \partial x_j}\right) \qquad (1.5)$$

Under certain simplifying assumptions, Eq. (1.5) reduces to

$$f(c, \nabla c, \nabla^2 c) = f_o(c) + k_1 \nabla^2 c + k_2 (\nabla c)^2 + \cdots \qquad (1.6)$$

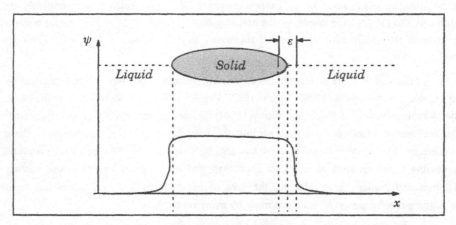

Figure 1.3. Distribution of the phase variable in the phase field model.

where k_1 and k_2 are appropriately chosen parameters. Thus, the free energy in a nonuniform system is a sum of a local free energy and a *gradient energy* contribution. The local Helmholtz free energy density, f, is postulated as a double-well function with two distinct minima, each corresponding to a phase, as shown in Fig. 1.4. Integrating over a volume V, we obtain the total free energy of the system from

$$F = N_v \int_V f dV \tag{1.7}$$

where N_v is the number of molecules per unit volume. At equilibrium one expects the functional F, as defined in Eq. (1.7), to be a minimum with respect to variations in ϕ, and thus ϕ satisfies the Euler-Lagrange equation

$$\frac{\delta F(\phi(\vec{x}))}{\delta \phi(\vec{x})} = 0 \tag{1.8}$$

This leads to

$$\frac{\delta F(\phi(\vec{x}))}{\delta \phi(\vec{x})} = f'(\phi(\vec{x})) - b\nabla^2\phi(\vec{x}) \tag{1.9}$$

where b is a constant. The kinetics of the function $\phi(x,t)$ usually is assumed to follow

$$\frac{\partial \phi(\vec{x}, t)}{\partial t} = -B\frac{\delta F(\phi_t(.))}{\delta \phi_t(\vec{x})} = B(b\nabla^2\phi_t(\vec{x}) - f'(\phi_t(\vec{x}))) \tag{1.10}$$

where B is always positive and may depend on ϕ and the temperature. Equation (1.10) implies that the free energy decreases along solution paths, that is the free energy is a Lyapunov functional. The equations solved then are the heat conduction equation with a source term accounting for latent heat evolution,

$$\frac{\partial T}{\partial t} - L\frac{\partial \phi}{\partial t} = \nabla^2 T \tag{1.11}$$

where L is the latent heat, and the equation for the evolution of the phase field

$$a\xi^2\phi_t = -\frac{\delta F}{\delta \phi} = \xi^2\nabla^2\phi + \frac{1}{2a}\phi(1 - \phi^2) + 2T \tag{1.12}$$

where $T=0$ is the phase transition temperature and a and ξ are constants. Here the Helmholtz energy functional used is

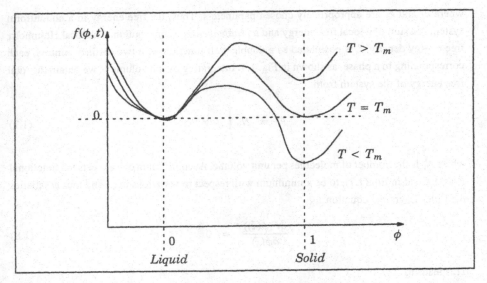

Figure 1.4. Free energy density f as a function of the phase fraction for different temperatures.

$$f(\phi, t) = \frac{1}{8a}(\phi^2 - 1)^2 - 2T\phi \qquad (1.13)$$

which displays the double-well functional behavior.

The solution of these equations leads to the development of realistic solidification patterns. The phase field model thus yields excellent qualitative results. However, in a general situation, the basis of the model is questionable. In a nonequilibrium situation the free energy may no longer be a Lyapunov functional. In fact, a minimization principle may not even hold (Langer 1980). Furthermore, it is difficult to correlate the several constants and parameters involved in the model to real physical systems. In addition, one has to balance the need for an extremely fine grid resolution to capture sharp interfaces with stability considerations for time-stepping. Thus, the phase field approach, while promising, cannot yet be regarded as a simulation technique for general solidification processes.

Recently, Roosen and Taylor (1994) have proposed a computational model for crystal growth, in which the interface between liquid and solid is exactly tracked, but the measurement of curvature is simplified through the assumption that the crystal is a polygon having a limited number of possible normal directions. These authors reported that their approach has the advantage that the computations involving the motion of the interface are relatively fast as compared to the above mentioned phase field algorithms; it is also relatively easy, compared to other interface tracking methods, to handle topological changes. They have presented interesting but limited results specialized toward crystal growth. This method has computational merits and appears worth pursuing further.

As far as the microstructural evolution is concerned, most computations mentioned thus far have been restricted to diffusion-controlled growth. Under this restriction, several of the numerical techniques mentioned above are applicable in solving the full governing equations or simplified models. For example, since the early 1980s, the physics of solidification instabilities have become more better understood as a result of local models for interface evolution (Brower et al. 1983). These models essentially amounted to computing the dynamics of strings under curvature-dependent motion, and no account was taken of the long-range behavior of the thermal field. Fixed-grid techniques have been applied to the solidification problem, with considerable success at the macroscopic scale. These techniques usually employ variants of the entropy-porosity formulation (Shyy and Rao 1994a, Voller and Prakash 1987). At the microscopic scales two useful methods are the aforementioned level set Hamilton-Jacobi formulation and the phase field models. A very comprehensive source of information on the theory and analysis of crystal growth can be found in the volumes edited by Hurle (1993). With regard to fluid flow characteristics observed in metals processing, Samanta et al. (1987), Szekely (1979), and Szekely et al. (1988) have collected some interesting information from materials engineers' viewpoint.

In the area of advanced materials processing techniques, many very interesting problems appear in the context of moving boundary problems. For example, Evans and Greif (1994) have made an attempt to model the chemical vapor deposition process of silicon nitride in a low–pressure hot–wall reactor, where they have also considered multicomponent diffusion along with chemical kinetics. Laser processing, another highly complicated technique, including hardening, drilling, melting, and quenching processes, has received much attention in the last two decades or so; for an overview and related literature, see Mazumder (1991). Representative efforts in formulating laser and phase change interaction can be found in Chan and Mazumder (1987), Kar et al. (1992), Kar and Mazumder (1990), Koai et al. (1993), Kurz and Trivedi (1992), and Zweig (1991).

One of the basic problems encountered by any model attempting to simulate the physics involved in solidification is the wide disparity of length scales. The length scales important for solidification processes arise from: (1) capillary forces, (2) heat conduction, (3) solutal diffusion, and (4) convection. With regard to convection, different mechanisms, including forced convection, natural convection, and Marangoni convection can all contribute to its strength. As will be discussed in more detail later, a rich variety of patterns can result, including cells, fingers, and dendrites, depending on the parameters present in the system. It should be pointed out that the identical set of governing equations can be used to describe the dynamic interaction between two fluids with different viscosities and confined by a very small length scale, the so-called Taylor-Saffman instability in the Hele-Shaw flow (Lamb 1932, Maxworthy 1986, McLean and Saffman 1981, Pelce 1988 Saffman and Taylor 1958).

With regard to heat transfer involving phase change, solidification and melting processes will be described in detail in this book. Boiling and condensation phenomena also share the same needs with regard to tracking moving boundaries. Relevant references are abundant in the literature; in terms of formulation and computation, the reader can refer to Carey (1992), Clift et al. (1978), Filipovic et al. (1994), Kaviany (1994), Klausner et al. (1993), Lee and Nydahl (1989), Levich (1962), and Mei et al. (1995) for more details.

Another area of substantial practical interest involving moving boundaries is the above-mentioned Hele-Shaw flow (Yih 1979). This problem is relevant to, for example, oil reservoir applications. It also has many interesting features and patterns resulting from the interaction between pressure and capillarity. For example, the motion of bubbles in a Hele-Shaw flow can exhibit remarkable shapes and unexpected translational velocities (Kopf-Sill and Homsy 1988, Maxworthy 1986); even the steady-state bubble shape may deviate significantly from the static circular shape. This problem also has been studied computationally by many researchers, such as DeGregoria and Schwartz (1986) using the boundary integral method, and Meiburg and Homsy (1988) using the Lagrangian vortex method. Even though this problem has been actively researched for more than 35 years, some important aspects of it remain inadequately understood, both mathematically and physically (Saffman 1986). Other more complicated physical features also have been added to investigate this problem in a broader context. For example, the effect of surfactant on interfacial characteristics (Rosen 1989) has been investigated recently by Park et al. (1994). Interesting features common to solidification processes and two-fluid systems such as Hele-Shaw flow and viscous fingering have been reported; together, they help provide better understanding of the moving boundary problem in general. Much of this information is also relevant to transport dynamics in porous media (Bejan 1984, Bear 1988, Kaviany 1991).

In addition to the above mentioned techniques, other approaches have been employed under some more specialized circumstances. Stone and Leal (1989) have applied a boundary integral method to compute the deformation, relaxation, and breakup processes of a viscous drop in a nonuniform velocity field. Their technique, which is based on that developed by Rallison and Acrivos (1978), is predicated on the assumption that the inertia of the fluid flow can be neglected. Tran-Son-Tay et al. (1991) have utilized the variational method, aided by the general solutions to the axisymmetric creeping flow equations, to simulate the dynamics of a highly deformed viscous drop as it reverts back to its resting spherical shape under the influence of surface tension. Relevant reviews of droplet and bubble dynamics and associated numerical techniques can be found in Rallison (1984) and in Stone (1994). Also in the area of biofluid dynamics, nonlinear instabilities resulting from the interaction of fluid and elastic characteristics of a thin film coating the inner surface airways of the lungs have been analyzed by Halpern and Grotberg (1992). Nonlinear evolution equations for the film thickness and wall position are derived in that work. In the highly convective regime, Lee and Chiu (1992) have used the pressure-based algorithm, summarized in the next chapter, to study physiological flows in an aortic bifurcation. Cheer and van Dam (1993) have reviewed some interesting papers in this general area.

1.2.3 Summary

In this chapter, we first outlined the objective and structure of this work. We subsequently presented the various methods currently available for tracking highly distorted fronts and outlined their advantages and limitations. This knowledge will guide us in designing the main features of the computational techniques to be developed in subsequent chapters. It is appropriate to emphasize that the particular approach taken to track moving boundaries is problem dependent. The physical problem dictates the complexity of the physics, in terms of both the interface and the flow behavior. Depending on the problem, numerical techniques with varying levels of complexity have to be developed. The level of sophistication desired is motivated by accuracy requirements in dealing with the interface, particularly if the interface shape becomes highly convoluted. In addition, the interaction of the interface and the flowfield also may determine the choice. In some cases, as will be evident in the examples to follow, the intricate details of the interface features may not be so important for predicting the flow behavior, while in others, it may in fact be the crucial feature. Obviously, one also has to balance the need for accuracy and physical realizability with the efficiency and ease of formulation and implementation of the chosen numerical scheme.

We wish to demonstrate in the following chapters the nuances and variations in approach that can be incorporated in solving a variety of physical problems. Although the use of well-tested, robust numerical algorithms is always preferable, for many problems it may be necessary to come up with innovative approaches. Consequently, the numerical solution of moving boundary problems is still an open area of research.

GOVERNING EQUATIONS AND SOLUTION PROCEDURE

In this chapter the fundamental fluid dynamic conservation laws are presented for a Newtonian fluid. The governing equations are transformed to a moving, body-fitted coordinate system using generalized curvilinear coordinates (Thompson et al. 1985). A brief description of a pressure-based algorithm suitable for the solution of the Navier-Stokes and associated transport equations is then given. The basic algorithm closely follows the original work of Patankar (1980) in Cartesian coordinates and Shyy (1994) in curvilinear coordinates. The moving grid treatment relies heavily on the notion of geometric conservation as described in Shyy (1994), Shyy and Vu (1991), Thomas and Lombard (1979), and Vinokur (1989). The algorithm presented in this chapter is one of a number of techniques useful for computing fluid flow problems, such as the artificial compressibility method (Chorin 1967, Hirsch 1990, Hosangadi et al., 1990, Kwak et al. 1986, Shuen et al. 1993), the penalty formulation (Braaten and Shyy 1986a, Hughes et al. 1979, Temam 1978), the projection method (Bell et al. 1989, Chorin 1968, Dukowicz and Dvinsky 1992), and the pressure–implicit with operator splitting (PISO) algorithm (Chen et al. 1992, Issa 1985).

In the context of the pressure-correction method, besides the materials presented in the following, the reader should refer to Rhie and Chow (1983), Majumdar (1988), and Melaaen (1992) for discussions on the application of nonstaggered and staggered grid layouts for the flow variables. General references can be found in some other comprehensive sources of information, such as Anderson et al. (1984), Canuto et al. (1988), Gottlieb and Orszag (1977), Fletcher (1988), Hirsch (1988, 1990), Jaluria and Torrance (1986), Minkowycz et al. (1988), Peyret and

Taylor (1983), Roache (1972), and Shyy (1994). Useful information related to solution techniques for fluid flows involving complex physics and geometries, such as internal boundaries, can be found in Oran and Boris (1987). Besides finite volume/difference methods, which are our focus here, spectral methods (Canuto et al. 1988, Orszag 1980, Patera 1984) and finite element methods (Baker 1983, Carey and Oden 1986, Hughes 1987, Zienkiewicz and Taylor 1991) have been extensively developed to handle complex geometries as well.

The following algorithm is chosen for presentation because it is used to obtain the solutions of the cases and examples discussed in this book, and because it has proven to be naturally compatible with the moving boundary techniques discussed. Wherever possible, we will adopt a consistent notation between the pressure-based algorithm and moving boundary techniques to help streamline the presentation. Besides the algorithm development aspect, several examples are presented to illustrate the issues involved in the implementation of a pressure correction method in curvilinear coordinates with a time dependent grid system and the treatment of a free surface based on the Young–Laplace equation.

2.1 FORMULATION

2.1.1 Governing Equations

The governing equations in Cartesian coordinates for two-dimensional, incompressible flow can be written in dimensional form as:

continuity:
$$\frac{\partial \varrho}{\partial t} + \frac{\partial (\varrho u)}{\partial x} + \frac{\partial (\varrho v)}{\partial y} = 0 \tag{2.1a}$$

x-momentum:
$$\frac{\partial (\varrho u)}{\partial t} + \frac{\partial (\varrho u u)}{\partial x} + \frac{\partial (\varrho u v)}{\partial y} = -\frac{\partial p}{\partial x} + \left[\frac{\partial}{\partial x} \left(\mu \frac{\partial u}{\partial x} \right) + \frac{\partial}{\partial y} \left(\mu \frac{\partial u}{\partial y} \right) \right] \\ + S_u(x, y) \tag{2.1b}$$

y-momentum:
$$\frac{\partial (\varrho v)}{\partial t} + \frac{\partial (\varrho u v)}{\partial x} + \frac{\partial (\varrho v v)}{\partial y} = -\frac{\partial p}{\partial y} + \left[\frac{\partial}{\partial x} \left(\mu \frac{\partial v}{\partial x} \right) + \frac{\partial}{\partial y} \left(\mu \frac{\partial v}{\partial y} \right) \right] \\ - \left(\varrho - \varrho_{ref} \right) g + S_v(x, y) \tag{2.1c}$$

energy:
$$\frac{\partial (\varrho C_p T)}{\partial t} + \frac{\partial (\varrho u C_p T)}{\partial x} + \frac{\partial (\varrho v C_p T)}{\partial y} = \left[\frac{\partial}{\partial x} \left(k \frac{\partial T}{\partial x} \right) + \frac{\partial}{\partial y} \left(k \frac{\partial T}{\partial y} \right) \right] \\ + S_T(x, y) \tag{2.1d}$$

In the above, ϱ is the density, u is the velocity component in the x–direction, v, is the velocity component in the y–direction, p is the pressure, T is the temperature, μ is the viscosity, k is the conductivity, g is acceleration due to gravity, and C_p is the specific heat. Here, the gravity vector $\vec{g} = -g\,\hat{j}$ is assumed to be in the $-y$ direction. For inclined configurations, the appropriate components should appear in both the momentum equations. S_u, S_v and S_T are appropriate source terms in the governing equations. The form of the source terms will be described later.

2.1.2 Governing Equations in a Body-Fitted Coordinate System

It frequently is convenient to write and solve the governing equations in generalized, nonorthogonal, curvilinear coordinates, especially when complex geometries are involved. Furthermore, in many applications it may be desirable to preferentially improve resolution in certain regions of the flow. These regions may be determined manually or may evolve dynamically during the solution process. The grid is then required to adapt to the regions of interest and the inclusion of moving coordinates is necessary. The continuity equation, (2.1a) in nonorthogonal coordinates, $[\xi = \xi(x, y, t), \ \eta = \eta(x, y, t)]$, then becomes,

$$\frac{\partial}{\partial t}(J\rho) + \frac{\partial(\rho U)}{\partial \xi} + \frac{\partial(\rho V)}{\partial \eta} = 0 \tag{2.2a}$$

where

$$J = x_\xi y_\eta - x_\eta y_\xi \tag{2.2b}$$

is the Jacobian of the transformation and

$$U = (u - \dot{x})\, y_\eta - (v - \dot{y})\, x_\eta$$
$$V = (v - \dot{y})\, x_\xi - (u - \dot{x})\, y_\xi \tag{2.2c}$$

are the contravariant velocity components. In Eq. (2.2c), $\dot{x}$ and $\dot{y}$ are the Cartesian components of the grid velocity vector. The momentum equations (2.1b) and (2.1c) then become

$$\frac{\partial(J\rho u)}{\partial t} + \frac{\partial(\rho U u)}{\partial \xi} + \frac{\partial(\rho V u)}{\partial \eta} = -\left\{ y_\eta \frac{\partial p}{\partial \xi} - y_\xi \frac{\partial p}{\partial \eta} \right\} + \frac{\partial}{\partial \xi}\left[\frac{\mu}{J}\left(q_1 u_\xi - q_2 u_\eta \right) \right]$$
$$+ \frac{\partial}{\partial \eta}\left[\frac{\mu}{J}\left(-q_2 u_\xi + q_3 u_\eta \right) \right] + S_u(\xi, \eta) \cdot J \tag{2.2d}$$

$$\frac{\partial(J\rho v)}{\partial t} + \frac{\partial(\rho U v)}{\partial \xi} + \frac{\partial(\rho V v)}{\partial \eta} = -\left\{ x_\xi \frac{\partial p}{\partial \eta} - x_\eta \frac{\partial p}{\partial \xi} \right\} + \frac{\partial}{\partial \xi}\left[\frac{\mu}{J}\left(q_1 v_\xi - q_2 v_\eta \right) \right]$$
$$+ \frac{\partial}{\partial \eta}\left[\frac{\mu}{J}\left(-q_2 v_\xi + q_3 v_\eta \right) \right] - \left(\rho - \rho_{ref} \right) g \cdot J + S_v(\xi, \eta) \cdot J \tag{2.2e}$$

23

along x and y directions respectively, where x_η, y_η, and so on are the metrics of the transformation and q_1, q_2 and q_3 are defined as

$$q_1 = x_\eta^2 + y_\eta^2$$
$$q_2 = x_\xi x_\eta + y_\xi y_\eta \qquad\qquad (2.2f)$$
$$q_3 = x_\xi^2 + y_\xi^2$$

The energy and other scalar transport equations can be written down in a similar manner and will not be repeated here. The advantage of this form is that complex geometries can be fitted by body-conforming grids, whereby boundary conditions can be applied easily and the grid can be distributed preferentially in desired regions so as to obtain better resolution.

2.2 DISCRETIZATION OF THE CONSERVATION LAWS

2.2.1 Pressure-Based Algorithm

A pressure-based numerical procedure originally proposed by Patankar and Spalding (1972) for Cartesian coordinates was chosen for computing laminar incompressible flows of interest in this work. The details of the basic pressure correction algorithm are given in Patankar (1980), with the extension of the procedure to general curvilinear coordinates given in Shyy (1994). The present implementation of the algorithm follows the work of Braaten and Shyy (1986b) and Shyy et al. (1985), with the statement of the conservation laws in terms of a time-dependent grid taken from Thompson et al. (1985).

A staggered grid arrangement is adopted for the discretization of Eqs. (2.1a–2.1d) as shown in Figure 2.1. Such a layout of the flow variables has been extensively adopted in computations of incompressible flows. Good stability and convergence characteristics have been obtained using a staggered grid for a wide range of complex flows (Shyy 1994). A nonorthogonal body-fitted grid system is generated numerically, with the nodes lying at the positions marked by triangles. Both the Cartesian and contravariant velocity components, as well as the Cartesian components of the grid velocity vector, are located at the midpoint of the control volume faces as shown in the figure. The pressure is located at the center of the four adjacent grid points. The finite-volume form of the conservation laws may be obtained by integrating Eqs. (2.1a–2.1d) using a first order, fully implicit time integration scheme over appropriately staggered control volumes leading to a five point difference stencil centered around the point P with neighbor points N, S, E, and W. By arbitrarily taking the grid spacing in the transformed domain to be unity and the time step to be Δt, the momentum and mass conservation laws take on the following discrete forms.

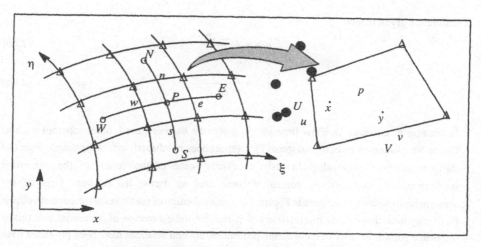

Figure 2.1 Staggered grid arrangement in the physical domain

continuity:

$$\frac{(J\varrho - J^0\varrho^0)}{\Delta t} + (\varrho U)_e - (\varrho U)_w + (\varrho V)_n - (\varrho V)_s = 0 \qquad (2.3)$$

u–momentum:

$$\frac{(J\varrho u - J^0\varrho^0 u^0)}{\Delta t} + \left(\varrho U u - \frac{\mu}{J}(q_1\frac{\partial u}{\partial \xi} - q_2\frac{\partial u}{\partial \eta}) + \frac{\partial y}{\partial \eta}p \right)_e$$

$$- \left(\varrho U u - \frac{\mu}{J}(q_1\frac{\partial u}{\partial \xi} - q_2\frac{\partial u}{\partial \eta}) + \frac{\partial y}{\partial \eta}p \right)_w$$

$$+ \left(\varrho V u - \frac{\mu}{J}(-q_2\frac{\partial u}{\partial \xi} + q_3\frac{\partial u}{\partial \eta}) - \frac{\partial y}{\partial \xi}p \right)_n$$

$$- \left(\varrho V u - \frac{\mu}{J}(-q_2\frac{\partial u}{\partial \xi} + q_3\frac{\partial u}{\partial \eta}) - \frac{\partial y}{\partial \xi}p \right)_s = 0 \qquad (2.4)$$

v–momentum:

$$\frac{(J\varrho v - J^0\varrho^0 v^0)}{\Delta t} + \left(\varrho U v - \frac{\mu}{J}(q_1\frac{\partial v}{\partial \xi} - q_2\frac{\partial v}{\partial \eta}) + \frac{\partial x}{\partial \eta}p \right)_e$$

$$- \left(\varrho U v - \frac{\mu}{J}(q_1\frac{\partial v}{\partial \xi} - q_2\frac{\partial v}{\partial \eta}) + \frac{\partial x}{\partial \eta}p \right)_w$$

$$+ \left(\varrho V v - \frac{\mu}{J}(-q_2\frac{\partial v}{\partial \xi} + q_3\frac{\partial v}{\partial \eta}) - \frac{\partial x}{\partial \xi}p \right)_n$$

$$- \left(\varrho V v - \frac{\mu}{J}(-q_2\frac{\partial v}{\partial \xi} + q_3\frac{\partial v}{\partial \eta}) - \frac{\partial x}{\partial \xi}p \right)_s = 0 \qquad (2.5)$$

where the superscript ^{0} refers to the previous time level. The Cartesian components of the grid velocity vector appearing in Eqs. (2.2c) are approximated by the first order backward time difference given below:.

$$\dot{x} = \frac{x - x^0}{\Delta t} \tag{2.6a}$$

$$\dot{y} = \frac{y - y^0}{\Delta t} \tag{2.6b}$$

In the above, although an Euler time-stepping scheme has been used, other schemes such as Crank-Nicolson may be easily adopted. This aspect does not directly affect the development of the pressure-based algorithm. In order to clarify some of the details of the numerical implementation, consider the control volume used to derive the discrete form of the u-momentum equation as shown in Figure 2.2. The evaluation of the fluxes at the control volume faces may be achieved by various means. In particular, the estimation of the convective flux is important for the stability of the solution procedure. Several schemes have been proposed over the years to compute these fluxes (Shyy 1994); they differ in the manner in which the cell face velocity is obtained from the neighboring grid points. The interpolation or, in other words, the shape function chosen determines the order of accuracy of the discretization. For example, if the convective and the diffusive momentum fluxes are approximated using second-order central differences, the discrete form of the u-momentum equation may be written as

$$a_P u_P = a_E u_E + a_W u_W + a_N u_N + a_S u_S + S_u \tag{2.7a}$$

where, after explicitly imposing mass conservation, the coefficients in Eq. (2.7a) are given by

$$a_E = -\frac{1}{2}\varrho U|_e + \frac{\mu}{J} q_1|_e \tag{2.7b}$$

$$a_W = \frac{1}{2}\varrho U|_w + \frac{\mu}{J} q_1|_w \tag{2.7c}$$

$$a_N = -\frac{1}{2}\varrho V|_n + \frac{\mu}{J} q_3|_n \tag{2.7d}$$

$$a_S = \frac{1}{2}\varrho V|_s + \frac{\mu}{J} q_3|_s \tag{2.7e}$$

$$a_P = a_E + a_W + a_N + a_S + \frac{J^0 \varrho^0}{\Delta t} \tag{2.7f}$$

This notation is standard in the literature concerned with the SIMPLE algorithm (Patankar 1980).

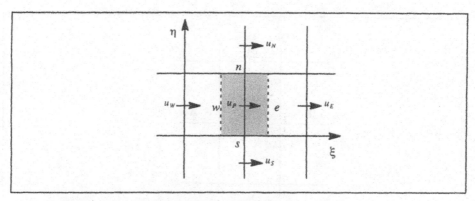

Figure 2.2   Control volume used for u-momentum equation.

The source term is given by

$$S_u = -\frac{\mu}{J}q_2\frac{\partial u}{\partial \eta}\Big|_e + \frac{\mu}{J}q_2\frac{\partial u}{\partial \eta}\Big|_w - \frac{\mu}{J}q_2\frac{\partial u}{\partial \xi}\Big|_n + \frac{\mu}{J}q_2\frac{\partial u}{\partial \xi}\Big|_s$$
$$+ \frac{\partial y}{\partial \eta}p\Big|_w - \frac{\partial y}{\partial \eta}p\Big|_e + \frac{\partial y}{\partial \xi}p\Big|_n - \frac{\partial y}{\partial \xi}p\Big|_s + \frac{J^0\varrho^0 u_P^0}{\Delta t} \qquad (2.7g)$$

where the various terms may be evaluated at the control volume faces using appropriate interpolations. In the present work both central and second-order upwind schemes (Shyy 1994, Shyy et al. 1992a) are used. When the second-order upwind scheme is used, additional terms appear in Eq. (2.7g).

Similarly, the discrete form of the v. momentum equation may be derived by considering the control volume shown in Figure 2.3. The discrete form of the equation may be written as

$$a_P v_P = a_E v_E + a_W v_W + a_N v_N + a_S v_S + S_v \qquad (2.8a)$$

with the coefficients, for a second-order central difference scheme, given by

$$a_E = -\frac{1}{2}\varrho U\Big|_e + \frac{\mu}{J}q_1\Big|_e \qquad (2.8b)$$

$$a_W = \frac{1}{2}\varrho U\Big|_w + \frac{\mu}{J}q_1\Big|_w \qquad (2.8c)$$

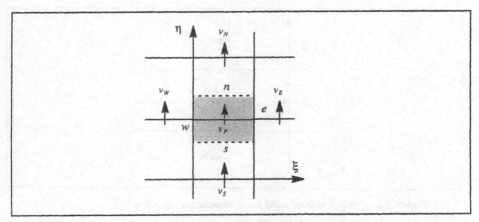

Figure 2.3. Control volume used for v-momentum equation

$$a_N = -\frac{1}{2}\varrho V|_n + \frac{\mu}{J}q_3|_n \qquad (2.8d)$$

$$a_S = \frac{1}{2}\varrho V|_s + \frac{\mu}{J}q_3|_s \qquad (2.8e)$$

$$a_P = a_E + a_W + a_N + a_S + \frac{J^0\varrho^0}{\Delta t} \qquad (2.8f)$$

and the source term given by

$$
S_v = -\frac{\mu}{J}q_2\frac{\partial v}{\partial \eta}\Big|_e + \frac{\mu}{J}q_2\frac{\partial v}{\partial \eta}\Big|_w - \frac{\mu}{J}q_2\frac{\partial v}{\partial \xi}\Big|_n + \frac{\mu}{J}q_2\frac{\partial v}{\partial \xi}\Big|_s
$$
$$
- \frac{\partial x}{\partial \eta}p\Big|_w + \frac{\partial x}{\partial \eta}p\Big|_e - \frac{\partial x}{\partial \xi}p\Big|_n + \frac{\partial x}{\partial \xi}p\Big|_s + \frac{J^0\varrho^0 v_P^0}{\Delta t} \qquad (2.8g)
$$

The discrete form of the pressure correction equation may be derived by considering the control volume shown in Fig. 2.4.

The discrete form of the pressure correction equation may be written as:

$$a_P\, p'_P = a_E\, p'_E + a_W\, p'_W + a_N\, p'_N + a_S\, p'_S + S_{p'} \qquad (2.9a)$$

with the coefficients given by

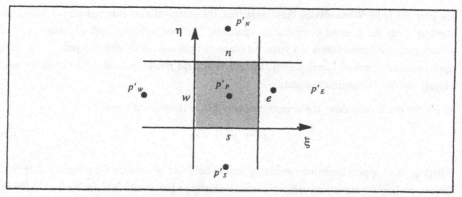

Figure 2.4. Control volume used for the pressure correction equation.

$$a_E = \varrho \left(\frac{y_\eta^2}{a_{u_P}} + \frac{x_\eta^2}{a_{v_P}} \right)_e \qquad (2.9b)$$

$$a_W = \varrho \left(\frac{y_\eta^2}{a_{u_P}} + \frac{x_\eta^2}{a_{v_P}} \right)_w \qquad (2.9c)$$

$$a_N = \varrho \left[\frac{y_\xi^2}{a_{u_P}} + \frac{x_\xi^2}{a_{v_P}} \right]_n \qquad (2.9d)$$

$$a_S = \varrho \left[\frac{y_\xi^2}{a_{u_P}} + \frac{x_\xi^2}{a_{v_P}} \right]_s \qquad (2.9e)$$

$$a_P = a_E + a_W + a_N + a_S \qquad (2.9f)$$

and the source term given by

$$S_{p'} = \varrho \left(U_w^* - U_e^* + V_s^* - V_n^* \right) + \frac{J^0 \varrho^0 - J\varrho}{\Delta t} \qquad (2.9g)$$

where the starred velocities denote the intermediate solution obtained by solving the momentum equations. In Eq. (2.9g), the right-hand side is the mass deficit that must be nullified in order for the velocity field to satisfy the continuity equation.

The sequence of events in the solution procedure is as follows. The momentum equations are first solved using an iterative method with an initial guess for the Cartesian velocities and the pressure field. When solving these equations, the contravariant velocities are first computed directly from the guessed Cartesian components. The new velocity field generally will not satisfy the mass conservation law since the pressure field used in the solution process was only approximately correct. Consequently, the pressure field must be corrected in order to more closely satisfy the continuity equation.

In the SIMPLE procedure, the correct pressure field p is obtained from

$$p = p^* + p'$$ (2.10a)

where p^* is an approximate intermediate pressure field and p' is called the pressure correction. Corresponding contravariant velocity corrections may be introduced in a similar way following the procedure described in Braaten and Shyy (1986b) and Shyy (1994).

$$U = U^* + U'$$ (2.10b)

$$V = V^* + V'$$ (2.10c)

To derive the pressure correction equation, we first subtract the momentum equations with the approximate pressure and velocity fields from the same momentum equations using the correct, continuity satisfying flow field variables. The resulting momentum correction equations then are assumed to be adequately approximated by the following truncated form

$$u'_p \, a_{u_p} \approx - \frac{\partial y}{\partial \eta} \frac{\partial p'}{\partial \xi} + \frac{\partial y}{\partial \xi} \frac{\partial p'}{\partial \eta}$$ (2.11a)

$$v'_p \, a_{v_p} \approx \frac{\partial x}{\partial \eta} \frac{\partial p'}{\partial \xi} - \frac{\partial x}{\partial \xi} \frac{\partial p'}{\partial \eta}$$ (2.11b)

Using these approximate correction equations, the Cartesian velocity components may be corrected by the following relationships

$$u = u^* + \left(- \frac{\partial y}{\partial \eta} \frac{\partial p'}{\partial \xi} + \frac{\partial y}{\partial \xi} \frac{\partial p'}{\partial \eta} \right) \Big/ a_{u_p}$$ (2.12a)

$$v = v^* + \left(\frac{\partial x}{\partial \eta} \frac{\partial p'}{\partial \xi} - \frac{\partial x}{\partial \xi} \frac{\partial p'}{\partial \eta} \right) \Big/ a_{v_p}$$ (2.12b)

Subsequently, the contravariant component correction formulas may be obtained by substituting Eq. (2.12) into Eq. (2.2c). After dropping terms that are not representable on a five point finite-difference molecule, the following simplified correction equations for the contravariant velocity components are obtained:

$$U = U^* - \left(\frac{y_\eta^2}{a_{u_P}} + \frac{x_\eta^2}{a_{v_P}} \right) \frac{\partial p'}{\partial \xi} \tag{2.13a}$$

$$V = V^* - \left[\frac{y_\xi^2}{a_{u_P}} + \frac{x_\xi^2}{a_{v_P}} \right] \frac{\partial p'}{\partial \eta} \tag{2.13b}$$

These equations then are substituted into Eq. (2.5) to obtain Eq. (2.9). The following residuals are defined to assess the degree of convergence of the discrete form of the fluid dynamic conservation laws.

$$R_u = \sum_{all\ cells} \left| \frac{- a_{up} u_p + \sum\limits_{all\ nb} a_u u_{nb} + S_u}{F_{ref}} \right| \tag{2.14a}$$

$$R_v = \sum_{all\ cells} \left| \frac{- a_{vp} v_p + \sum\limits_{all\ nb} a_v v_{nb} + S_v}{F_{ref}} \right| \tag{2.14b}$$

$$R_{mass} = \sum_{all\ cells} \left| \frac{\varrho U_e - \varrho U_w + \varrho V_n - \varrho V_s}{M_{ref}} \right| \tag{2.14c}$$

Here, the residuals are normalized by reference momentum and mass fluxes, F_{ref} and M_{ref}, respectively.

In the present procedure, the pressure corrections are used to update the contravariant components using Eq. (2.13). Provided the pressure correction equation is solved to convergence at each iteration, the resulting contravariant velocity fields exactly satisfy mass conservation. After satisfying continuity in terms of the contravariant components, it is necessary to recover the Cartesian velocity components before returning to the momentum equations with the updated velocity and pressure fields. It is of the utmost importance to consistently recover the Cartesian velocity components from the updated contravariant components since, from Eq. (2.5), mass conservation is stated explicitly in terms of the contravariant components, in generalized curvilinear coordinates. Due to the staggering of the velocity components care must be taken to ensure a consistent transformation between the

Cartesian components and contravariant components. If the transformation is done naively, an inconsistency will arise in the finite volume form of the conservation laws. This issue is briefly discussed in the following.

The obvious way to compute the Cartesian components from the contravariant components is to analytically invert the transformation given by Eq. (2.2c). This inversion gives the following formulas for the Cartesian components in terms of the contravariant velocity components:

$$u = \left(\frac{U}{J} \frac{\partial x}{\partial \xi} + \frac{V}{J} \frac{\partial x}{\partial \eta} \right) + \dot{x} \qquad (2.15a)$$

$$v = \left(\frac{V}{J} \frac{\partial y}{\partial \eta} + \frac{U}{J} \frac{\partial y}{\partial \xi} \right) + \dot{y} \qquad (2.15b)$$

In order to estimate the contravariant velocity components, it is necessary to interpolate for the value of v when computing U in Eq. (2.2c), and similarly to interpolate for the value of u when computing V. A corresponding interpolation is needed when computing u and v in Eq. (2.15). The need for these interpolations causes an inconsistency in mass conservation to arise in the procedure. An efficient method for consistently recovering the Cartesian components from the updated contravariant components based on the so-called D'yakonov iteration procedure is described in Braaten and Shyy (1986b) and Shyy (1994). In the present numerical procedure, the method developed in these references for computing the Cartesian velocity components is used.

2.2.2 Consistent Estimation of the Metric Terms

Due to the need for interpolation in computing the contravariant velocity components appearing in Eq. (2.15), another inconsistency may arise in the estimation of the metric terms. For example, consider the continuity equation when the grid is time-dependent. An inconsistent numerical implementation of the continuity equation would lead to the generation of artificial mass sources in the flow calculation. As suggested by Thompson et al. (1985), the following equation relating the time rate of change of the Jacobian to the Cartesian components of the grid velocity is used to update the Jacobian at the implicit time level:

$$\frac{\partial J}{\partial t} + \frac{\partial}{\partial \xi}(-\dot{x} \frac{\partial y}{\partial \eta} + \dot{y} \frac{\partial x}{\partial \eta}) + \frac{\partial}{\partial \eta}(-\dot{y} \frac{\partial x}{\partial \xi} + \dot{x} \frac{\partial y}{\partial \xi}) = 0 \qquad (2.16)$$

This identity may be derived from Eq. (2.5) by requiring mass to be conserved for a constant density, uniform velocity field under a time dependent coordinate transformation.

Integrating Eq. (2.16) using a first order, fully implicit time integration scheme over the same control volume used for mass conservation leads to Eq. (2.17), which is the finite volume, discrete form of the identity given above. Adopting this equation as the updating formula for the Jacobian guarantees that the basic requirement of geometric conservation (Shyy and Vu 1991, Thomas and Lombard 1979, and Vinokur 1989) is respected in the discrete form of the conservation laws when the grid is time dependent.

$$\frac{(J - J^0)}{\Delta t} + (- \dot{x} \; \frac{\partial y}{\partial \eta} + \dot{y} \; \frac{\partial x}{\partial \eta})_e - (- \dot{x} \; \frac{\partial y}{\partial \eta} + \dot{y} \; \frac{\partial x}{\partial \eta})_w$$
$$+ (- \dot{y} \; \frac{\partial x}{\partial \xi} + \dot{x} \; \frac{\partial y}{\partial \xi})_n - (- \dot{y} \; \frac{\partial x}{\partial \xi} + \dot{x} \; \frac{\partial y}{\partial \xi})_s = 0 \qquad (2.17)$$

In Eq. (2.17), the value of the Jacobian at the new time instant is evaluated according to the given grid and boundary velocities; conversely, one can estimate the Jacobian based on the instantaneous grid distribution, and, utilizing Eq. (2.17), estimate the grid and boundary velocity vectors accordingly (Bartina 1989, Demirdzic and Peric 1990, Lu et al. 1995). Apart from the Jacobian, it also is well established that, for example, precise geometric interpretation of the metric terms exist in the definition of the contravariant velocity components in Eq. (2.2c). This aspect is discussed in detail by Shyy (1994). These terms, such as x_η, y_η in Eq. (2.2c), must be evaluated as the projected lengths of the given cell face along x and y directions, respectively.

2.2.3   Illustrative Test Cases

Based on the discretization of the governing equations summarized above, we now present the following test cases to gain insight into some of the points concerning D'yakonov iteration and the consistent evaluation of metrics. These cases are useful test problems for establishing the reliability and accuracy of a pressure-based algorithm under various limiting conditions.

2.2.3.1   Rotated Channel Flow

The first test case is that of steady, developing parallel flow in a channel that is rotated 45° to the Cartesian axes. Once the flow becomes fully developed, the convection terms vanish in the Navier–Stokes equations and the familiar parabolic velocity profile solution with constant pressure gradient is obtained (Schlichting 1979).

Figure 2.5 shows the constant pressure contours computed using the numerical algorithm described in the previous chapter, with the D'yakonov iterative procedure used to recover the Cartesian velocity components from the contravariant components. A uniform velocity field is specified at the channel inlet and a zeroth order extrapolation of the Cartesian velocity components is used at the channel exit to enforce a gradient-free boundary condition in the

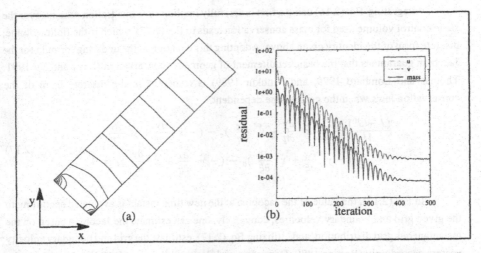

Figure 2.5. (a) Pressure contours and (b) convergence path for a 45° rotated channel at Re = 40. The results shown are for a 41 x 41 uniform grid. D'yakonov iteration was used in the computation.

streamwise direction. As seen in the figure, the streamwise pressure gradient becomes constant after some initial adjustment of the velocity and pressure fields near the inlet. A comparison of the computed streamwise pressure gradient with the analytic solution for fully developed parallel flow is shown in Fig. 2.6. Once the flow becomes fully developed, the computed pressure gradient is indistinguishable from the analytic solution, as expected. The residuals of the discrete form of the conservation laws, defined by Eq. (2.14), also are shown in the figure. The momentum equations converge to a terminal level of 5 x 10^{-4}, and the continuity equation converges to a terminal level of 5 x 10^{-5}. These levels of residuals are consistent with the single-precision floating point accuracy of the arithmetic used in the calculation.

Figure 2.7 shows the results for the same problem without employing the D'yakonov procedure; instead, the Cartesian velocity components are computed directly from Eq. (2.15). Interestingly, the pressure contours look qualitatively correct, but the momentum equation residuals have converged to a terminal level which is of order one. The reason for the lack of convergence of the momentum equations is evident in Fig. 2.6, where it may be seen that the pressure gradient is computed incorrectly when the D'yakonov iterative procedure is not used. The convergence path shown in Fig. 2.7 clearly illustrates the inconsistency that arises in the transformation due to the noncollocated velocity components. Fortunately, the convergence path

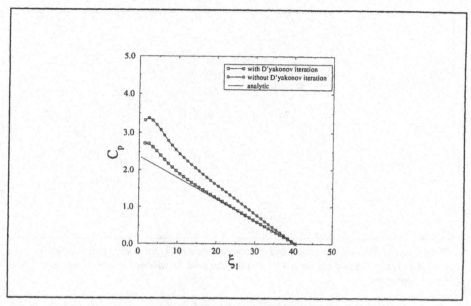

Figure 2.6. Centerline pressure coefficient for the 45^0 rotated channel at Re = 40.

in Fig. 2.5 shows that the inconsistency is resolved using the D'yakonov iterative procedure during the course of the computation.

2.2.3.2   *Uniform Flow Using a Moving Grid*

The second elementary test case is the computation of a uniform flow on a grid that moves arbitrarily in space as a function of time. This is an important test case. Any formulation that is to be applied to unsteady flow problems using time-dependent body-fitted coordinates must admit a uniform flow field. Furthermore, this solution must hold identically when the grid coordinates are an arbitrary function of time.

Figure 2.8 shows the grid and pressure contours after one time step for a uniform flow field inclined at 45° to the Cartesian coordinate axes. The initial grid at time t is shown in Fig. 2.8(a) and the grid at time $t + \Delta t$ is shown in Fig. 2.8(b). The boundary conditions imposed on the solution at all computational boundaries are Dirichlet conditions on the Cartesian velocity components. The pressure contours computed using Eq. (2.17) to update the Jacobian of the transformation at the $t + \Delta t$ time level are shown in Fig. 2.8(c). The computed pressure gradient in the flow is approximately 10^{-7} everywhere in the domain and the contours shown in the figure

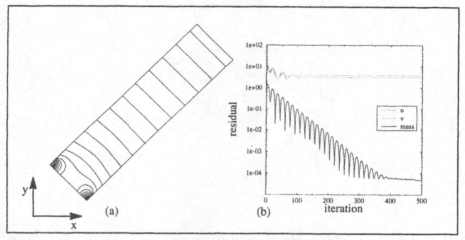

Figure 2.7 (a) Pressure contours and (b) convergence path, for a 45^0 rotated channel at Re = 40. The results shown are for a 41 x 41 uniform grid. D'yakonov iteration was not used in the computation.

result from roundoff error at the limit of single-precision floating point arithmetic. In contrast, the pressure contours computed using Eq. (2.2b) to calculate the Jacobian at the $t + \Delta t$ time level are shown in Fig. 2.8 (d). The pressure gradient in this case is order one, which clearly demonstrates the inconsistency that arises in the computation simply by taking a snapshot of the grid at the implicit time level and computing the Jacobian using Eq. (2.2b). Adopting Eq. (2.17) as the updating formula for the Jacobian guarantees that the basic conservation law is satisfied for unsteady flows using time-dependent body-fitted coordinates.

2.3 FORMULATION AND SOLUTION OF FLOWS WITH FREE SURFACES

2.3.1 Introduction

In addition to the algorithms appropriate for solving the fluid flow equations on stationary and moving grids, the formulation and solution procedure for free and internal boundaries needs to be addressed. The physical laws governing the formation and movement of the internal boundary will be included in the next two chapters, in the context of a fluid-membrane interaction problem and solidification problems. In the following, we present some important issues related to free boundary problems. To illustrate the features, a float zone with a meniscus separating the liquid and gas phases will be used as the primary example.

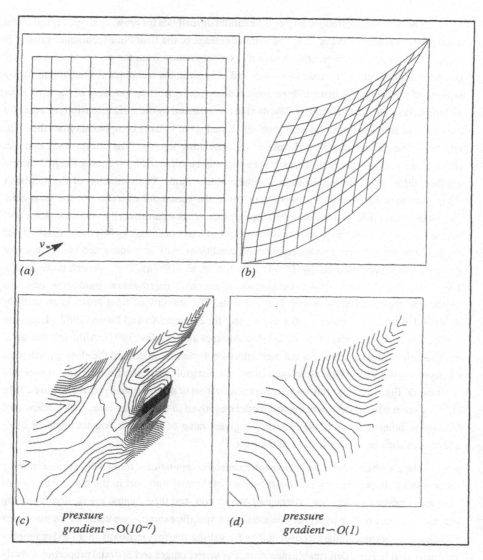

Figure 2.8. Comparison of Jacobian updating strategies for the computation of a 45° uniform flow on a time dependent grid. Initial and t+∆t grids are shown in (a) and (b), respectively. Computed pressure contours using Eq. (2.17) shown in (c), and using Eq. (2.2c) shown in (d).

The float zone technique is one of the candidate methods for growing single crystals. The configuration is shown in Fig. 2.9. The main advantage of the float-zone technique is that it is a containerless processing technique such that contamination from the wall–a main source of problems for crystal quality control–is removed. On earth, the technique is limited mainly by the size of the melt zone that can be achieved, since the hydrostatic pressure of the melt zone and the normal stresses induced by flow motion are balanced by the surface tension between the melt and the ambient fluid. Another problem is the loss of volatile components from the melt, which can be circumvented by the use of encapsulating liquids. The encapsulant also can increase the maximum size of the melt zone that can be achieved by reducing the hydrostatic pressure difference between the melt and the ambient fluid. The formation of the meniscus during the melting process is a major concern in the successful operation of the float-zone technique. This aspect and associated transport processes that can affect the meniscus shape will be the focus of the following sections. The issues of existence, multiplicity, and sensitivity of the liquid meniscii formed under equilibrium conditions, will be investigated based on a free energy concept presented in Shyy(1994) and Shyy et. al. (1993a). The approach taken here is static in nature, which is a satisfactory approximation considering the slow speed of the meniscus movement–typical of most practical applications. The stability of float zones is an actively investigated topic; for relevant information, see, for example Xu and Davis (1983), Langbein (1990, 1992), Sanz Andres (1992), and Slobozhanin and Perales (1993). Although the static consideration cannot account for the melt-interface dynamics, in general the slow growth rates of many crystal processing techniques allow meaningful application within this theoretical framework. Figure 2.9(a) shows a schematic illustration of the float-zone process. Figure 2.9(b) shows a schematic of the mathematical model employed in the present case, which allows the trijunction point to move so as to satisfy a given value of the static contact angle or other constraints such as a fixed volume of the melt.

Once we have established the theoretical and computational framework for determining the meniscus shape, we then address the issues of thermal transport in the melt. The role of buoyancy-induced convection, Marangoni convection, and their mutual interaction regarding thermal transport will be of primary concern. As a specific example, we consider the buoyancy induced and thermocapillary convection and resulting thermal transport in an axisymmetric geometry defined by a deformable meniscus. Parameter ranges and material properties directly applicable to an experimental configuration employed for the float-zone growth have been considered. Substantial effort has gone into ensuring solution accuracy and adequate grid resolution. Issues associated with the existence and uniqueness of the meniscus profiles are discussed. These issues have not been treated in previous studies. For simplicity, this study is restricted to convective thermal transport only. Considerations related to phase change will be addressed in later chapters.

2.3.2 Prediction of Meniscus Shapes

2.3.2.1 Methodology

Meniscus profiles obey the Young-Laplace equation in axisymmetric form (Finn 1986, Myshkis et al. 1987, Shyy 1994), which relates the free surface curvature to the pressure difference across it. The governing equation describes the generation of surface curvature to balance the sum of the hydrostatic pressure and applied/external pressurization. Specialized to an axisymmetric geometry and isothermal domain without melt convection, it can be written as follows:

$$\Delta\varrho \; g \; y - \Delta p = \gamma \left[\frac{f''}{(1 + f'^2)^{3/2}} - \frac{1}{f \, (1 + f'^2)^{1/2}} \right] \tag{2.18}$$

where $\Delta\varrho$ is the density difference between the ambient and the melt and Δp, also called the pressurization, is a source term which is determined as a part of the solution procedure by satisfying volume conservation. γ is the surface tension coefficient and the meniscus is described by $f = r(y)$, where r is the radius and y is the vertical coordinate. A variety of boundary conditions have been treated in the literature, such as fixed contact angles at both top and bottom (Concus 1990, Finn 1986, Myshkis et al. 1987), fixed radius at the bottom and fixed contact angle at the top (Shyy et al. 1993a) and fixed radius at both top and bottom (Coriell et al. 1977, Li et al. 1993, 1994, Zhang and Alexander 1992). These conditions have been devised for configurations under various physical constraints. For float zones, the imposed boundary conditions may vary depending on the operating conditions. Therefore, this study has been conducted with a variety of different constraints imposed on the solutions. The following boundary conditions may be imposed:

$$f \, (y{=}0) \; = r_b \text{ and}$$
$$\text{either } f \, (y{=}h_c) = r_c$$
$$\text{or } f'(y{=}h_c) \; = \tan(\pi/2 - \phi_c)$$

Figure 2.9(b) shows a schematic of the geometry. For an isothermal liquid bridge, Eq. (2.18) can be nondimensionalized with a length scale of r_b and a pressure scale of γ/r_b to yield,

$$\text{Bo } Y - \Delta P = \left[\frac{f'}{(1 + f'^2)^{3/2}} - \frac{1}{f \, (1 + f'^2)^{1/2}} \right] \tag{2.19}$$

where the italic f indicates the nondimensionalized quantity. Bo is the Bond number, which is

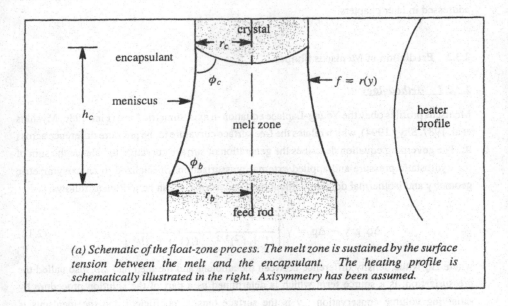

(a) Schematic of the float-zone process. The melt zone is sustained by the surface tension between the melt and the encapsulant. The heating profile is schematically illustrated in the right. Axisymmetry has been assumed.

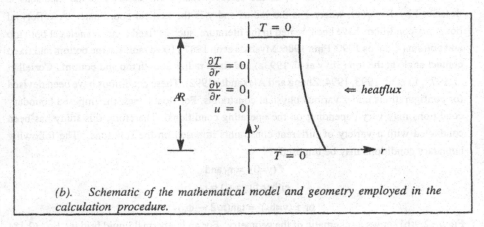

(b). Schematic of the mathematical model and geometry employed in the calculation procedure.

Figure 2.9.   Schematic of the float-zone process and the mathematical model employed.

defined as:

$$Bo = \frac{\Delta\varrho \, g \, r_b^2}{\gamma} \qquad (2.20)$$

In Shyy et al. (1993a), the nonlinear differential equation, Eq. (2.19), was solved subject to a fixed radius at the lower boundary and a fixed contact angle at the top boundary. Such a two point boundary value problem cannot yield an unique solution even for a linear differential equation. Thus, there arises the problem of selecting, out of the multiple solutions that are mathematically permissible, the one that corresponds to the physical equilibrium condition that exists in reality. Therefore, we now invoke the thermodynamic condition that, at equilibrium, the free energy is a minimum and select the solution that minimizes the free energy. For simplicity, we consider only the isothermal condition and examine the Helmholtz free energy, which contains three contributions:

(i)   The potential energy from the effective head.
(ii) The surface energy of the meniscus forming the gas-liquid interface.
(iii) The surface energy needed to wet the solid-liquid wetted area.

The equilibrium shape is the one that minimizes the total free energy:

$$E = \int\limits_{0}^{h_c} \left[\pi \, f^2 \, (\Delta\varrho g y - \Delta p) + 2\pi \, \gamma \, f \, (1 + f'^2)^{1/2} \right] dy - \pi \, \gamma \, r_c^2 \, \cos(\phi_0) \qquad (2.21)$$

where ϕ_0 is the static contact angle. The Young-Laplace equation may be integrated starting with a fixed radius and base angles, $-\pi/2 < \phi_b < \pi/2$, to yield all possible solutions for a given aspect ratio of the domain. Subsequently, solutions satisfying a given contact angle at the top, ϕ_c, are selected and their free energies calculated according to Eq. (2.21). If any of these solutions locally minimize the free energy, then these profile shapes are statically stable. As described in Shyy (1994) and Shyy et. al. (1993a) , the value of ϕ_0 is specified, and E is calculated by fixing the value of ϕ_0 in Eq. (2.21) and scanning through the whole range of meniscus profiles obtained by fixing the lower trijunction location (with varying angle) and the height of the upper trijunction point (with varying locations and angles). Distinct extrema are obtained at the specified values of ϕ_0. Among the multiple solutions, the one corresponding to a minimum on the curve is the physically realizable, statically stable meniscus profile. When either a maximum or a nonextremum point arises, we may surmise that such solutions belong to the unstable branch. Once the meniscus shape is obtained, the volume of the melt zone in nondimensional form is given by,

$$V = \int_0^{AR} \pi \, f^2 \, dy \tag{2.22}$$

where $AR = h_c/r_b$ is the aspect ratio of the zone. This constraint is important since the meniscus profiles are required to enclose a fixed volume of the melt satisfying Eq. (2.22) (Coriell et al. 1977, Li et al. 1993, Myshkis et al. 1987, Zhang and Alexander 1992). For example, in Zhang and Alexander (1992), the contact lines are fixed and the free surface shape is calculated by iteratively satisfying the governing equations and the volume constraint using a Picard type of iterative procedure. In Li et. al. (1993) and Lie et al. (1989), the meniscus contact angles are adjusted to iteratively satisfy the governing equations and the volume constraint. The basic considerations from free energy minimization and uniqueness can shed useful light on the issues relevant to the float zone. It also is noted that additional forces may be generated on the free surface due to electromagnetic effects if the zone is heated by an induction coil (Lie et al. 1989, Riahi and Walker 1989).

2.3.2.2 Effect of Convection on Meniscus Shape

It may be expected that convection in the melt will impact the meniscus shape through the normal stress terms which will be balanced by the free surface curvature. In addition, the free surface will no longer be an isotherm, which implies that the surface tension will vary from point to point depending on the spatial distribution of the temperature along the free surface. Under such conditions, Equation (2.18) must be modified as follows:

$$\Delta\varrho \, g \, y - \Delta p + 2\mu \frac{\partial v_n}{\partial n} = \gamma(T) \left[\frac{f''}{(1 + f'^2)^{3/2}} - \frac{1}{f \, (1 + f'^2)^{1/2}} \right] \tag{2.23}$$

where μ is the viscosity of the melt. This can be nondimensionalized in a manner similar to Eq. (2.18) along with the reference velocity scale defined as α/r_b, where α is the thermal diffusivity. In nondimensional form, this becomes:

$$\text{Bo } Y - \Delta P + \text{Ca} \frac{\partial V_n}{\partial n} = \left(1 - \left|\frac{dy}{dT}\right| \frac{\Delta T}{\sigma_o} \right) \left[\frac{f''}{(1 + f'^2)^{3/2}} - \frac{1}{f \, (1 + f'^2)^{1/2}} \right] \tag{2.24}$$

where Ca is the Capillary number defined as $(2 \, \mu \, a)/(\gamma_o \, r_b)$, γ_o is the surface tension at the reference temperature, and the italics denote the nondimensionalized variables. The Capillary number and the Bond number determine the relative influence of the hydrostatic and the convective effects, respectively. The term premultiplying the curvature term describes the correction due to the variation of surface tension with temperature. In this study, as will be presented later, it was found that the Capillary number, Ca, and the normal gradient of the normal velocity were both small compared to the hydrostatic term, from an a posteriori estimation, and hence meniscus shapes were generated based on Eq. (2.18). Only the steady state has been considered in this study.

2.3.3 Sources of Convection

2.3.3.1 Natural Convection

The strength of natural convection depends on the density differences caused by thermal and compositional inhomogeneities within the fluid. In two–dimensions, the source of buoyancy driven convection is the component of the local density gradient normal to the gravity vector. A temperature scale, ΔT, can be defined which, along with a characteristic length scale, r_b, characterizes the horizontal thermal gradient and consequently the horizontal density gradient. In addition, we can define a characteristic solute scale, $\Delta\phi$. The Boussinesq approximation is now invoked, which by linearization neglects the density variation in all the terms of the governing equations except the source term. The buoyancy source term then becomes

$$ -\left(\varrho - \varrho_{ref}\right) g = \varrho\, g \left[\beta_T \left(T - T_{ref}\right) + \beta_\phi \left(\phi - \phi_{ref}\right)\right] \qquad (2.25) $$

where β_T and β_ϕ are the coefficients of thermal expansion and solutal expansion, respectively, of the melt. The resulting nondimensional parameters are

Thermal Grashof number: $\qquad \mathrm{Gr}_T = \dfrac{g\,\beta_T\,\Delta T\, r_b^3}{\nu^2}$ $\qquad\qquad$ (2.26a)

Solutal Grashof number: $\qquad \mathrm{Gr}_\phi = \dfrac{g\,\beta_\phi\,\Delta\phi\, r_b^3}{\nu^2}$ $\qquad\qquad$ (2.26b)

Prandtl number: $\qquad\qquad\quad \mathrm{Pr} = \dfrac{\mu}{k/C_p}$ $\qquad\qquad$ (2.26c)

Schmidt number: $\qquad\qquad\quad \mathrm{Sc} = \dfrac{\nu}{D}$ $\qquad\qquad$ (2.26d)

where k is the thermal conductivity, ϱ is the density, C_p is the heat capacity, μ is the dynamic viscosity, ν is kinematic viscosity, and D is the mass diffusivity. The thermal and the solutal Rayleigh numbers can be defined as $\mathrm{Ra}_T = \mathrm{Gr}_T \cdot \mathrm{Pr}$ and $\mathrm{Ra}_\phi = \mathrm{Gr}_\phi \cdot \mathrm{Sc}$, respectively.

2.3.3.2 Marangoni Convection

For problems involving free or deformable interfaces between two liquids or between a liquid and a gas/vapor, temperature and composition inhomogeneities can result in a surface tension gradient along the interface. This exerts a shear stress on the fluid giving rise to Marangoni convection. The float zone is an example of a configuration where Marangoni convection plays an important role in governing the solidification characteristics of the melt. General discussion of Marangoni convection can be found in Koschmieder 1993, Shyy 1994). We now briefly examine the case of thermally generated Marangoni convection.

For liquids bounded by a free or deformable boundary where a temperature gradient exists along the interface, a surface shear stress is generated by the surface tension gradient. Thus, the force balance can be written as

$$\nabla \, (\vec{V} \cdot \vec{t}) \cdot \vec{n} = \frac{\partial \gamma}{\partial T} \, \nabla T \cdot \vec{t} \tag{2.27}$$

where the surface tension gradient is proportional to the temperature gradient and gives rise to the normal derivative of the tangential velocity at the free surface. $\vec{V}$ is the velocity vector, $\vec{n}$ is the unit normal vector to the free surface, and $\vec{t}$ is the unit tangent vector to the free surface. The convection strength is described, in nondimensional terms, by the Marangoni number,

$$\mathrm{Ma} = \left| \frac{d\gamma}{dT} \right| \frac{\Delta T \, r_b}{\mu \alpha} \tag{2.28}$$

This dimensionless parameter represents the balance between the shear stress and the surface tension gradient after the velocity and space variables have been nondimensionalized by the velocity scale based on heat conduction, as will be described in the next section.

2.3.4   Nondimensionalization and Scaling Procedure

The governing equations can be nondimensionalized with reference length, velocity and thermal scales. The choice of nondimensional parameters used to normalize the governing equations is not unique as will be apparent in the following.

The dimensionless variables can be defined as:

$$\tau = t/t_r \, , \quad X = x/d \, , \quad Y = y/d$$
$$\bar{u} = u/U_r \, , \quad \bar{v} = v/U_r$$
$$P = p/(\varrho U_r^2) \text{ and } \Theta = (T - T_c)/(T_h - T_c) \tag{2.29}$$

In (2.29), d is the reference length scale, U_r is the reference velocity scale and the temperature is normalized by the reference temperature scale, $\Delta T = T_h - T_c$, such that the nondimensional temperature is between zero and unity. The reference time scale is defined as $t_r = d/U_r$. The reference velocity scale is determined by the dominant physical processes occurring in the system as described next.

2.3.4.1 Heat Conduction Scales

The reference velocity scale for heat conduction is

$$U_r = \frac{\alpha}{d} \qquad (2.30)$$

where $\alpha = k/(\varrho C_p)$ is the thermal diffusivity of the fluid. Then, with the reference scales specified earlier, the non-dimensional equations become:

momentum: $\qquad \dfrac{\partial \vec{V}}{\partial \tau} + \vec{V} \cdot \nabla(\vec{V}) = -\nabla P + \Pr \nabla^2 \vec{V} + \operatorname{Ra}\Pr \Theta \hat{g} \qquad (2.31a)$

energy: $\qquad\qquad\qquad \dfrac{\partial \Theta}{\partial \tau} + \vec{V} \cdot \nabla\Theta = \nabla^2 \Theta \qquad (2.31b)$

where $\vec{V}$ is the nondimensional velocity vector, Pr is the Prandtl number and Ra is the thermal Rayleigh number.

2.3.4.2 Natural Convection Scales

The reference velocity scale for natural convection is

$$U_r = \sqrt{g\,\beta\,d\,\Delta T} \qquad (2.32)$$

and with the reference scales specified earlier, the nondimensional equations become

momentum: $\qquad \dfrac{\partial \vec{V}}{\partial \tau} + \vec{V} \cdot \nabla(\vec{V}) = -\nabla P + \dfrac{1}{\sqrt{\operatorname{Gr}_T}}\nabla^2 \vec{V} + \Theta \hat{g} \qquad (2.33a)$

energy: $\qquad\qquad\qquad \dfrac{\partial \Theta}{\partial \tau} + \vec{V} \cdot \nabla\Theta = \dfrac{1}{\Pr \sqrt{\operatorname{Gr}_T}}\nabla^2 \Theta \qquad (2.33b)$

where Pr is the Prandtl number and Gr_T is the thermal Grashof number.

2.3.4.3 The Marangoni Number

If equation (2.27) is nondimensionalized with respect to the heat conduction scales (2.30), described in section 2.3.4.1, the Marangoni number defined by equation (2.28) occurs naturally in the dimensionless form

$$\nabla (\vec{V} \cdot \vec{t}) \cdot \vec{n} = \operatorname{Ma} \nabla\Theta \cdot \vec{t} \qquad (2.34)$$

Considering that the dimensionless temperature gradient is of order one, the above equation indicates that the strength of the shear stress in the free surface region is proportional to the Marangoni number; this shear stress, in turn, induces convection within the body of the fluid. Thus, the Marangoni number describes the strength of convection due to the surface tension gradient.

2.3.5 Formulation and Computational Algorithm for Transport Processes

The most influential factors governing the crystal growth process are the heat and solute transport processes in the melt, which control the shape and movement of the solidifying front. In the following section, a meniscus configuration frequently encountered in practical float-zone processes is isolated for detailed analysis of heat transport within the melt. Of particular interest is the role of buoyancy-driven and thermocapillary convection processes and the interaction of these processes with the shape of the free surface in determining the heat transfer within the melt. Phase change and solute transport effects are presented here. The Navier-Stokes equations, along with the energy equation are written in cylindrical coordinates to facilitate the treatment of the axisymmetric geometry shown in Figure 2.9(b). The numerical scheme involves the discretization of the transformed form of the governing equations based on a control volume formulation as described in Shyy (1994), Shyy et al. (1985), and Shyy and Vu (1991) and outlined earlier in this chapter. The relevant aspects of this procedure are described below (also see Shyy and Rao 1994b). An appropriate choice of scales for the nondimensionalization procedure was arrived at by selecting the base radius r_b as the characteristic length scale and the thermal diffusion velocity scale, defined as α/r_b, where α is the thermal diffusivity. The selection of the temperature scale is not as straightforward because the boundary condition at the free surface is specified as a heat flux from which an estimate of the maximum temperature in the domain has to be derived. The procedure is explained below in detail. Since the heat flux is specified on the free surface, a temperature scale has to be derived in order to nondimensionalize the energy equation and to define the Grashof and the Marangoni numbers. It would seem natural to define a temperature scale based on the heat flux

$$\Delta T = \frac{q_{max}}{k} \, d \qquad (2.35a)$$

where q_{max} is the maximum value of the heat flux specified on the boundary. However, it has been observed that such a definition consistently overestimates the temperature scale, that is, the maximum temperatures achieved in the domain, even for conduction cases, are of the order of five times less than the value suggested by Eq. (2.35a). In addition, it is expected that the maximum temperatures obtained by taking convection into account should be even less. Therefore, the temperature scale obtained above will be scaled down by a factor of ten in order to obtain a more realistic temperature scale. Thus, the temperature scale used to calculate the Grashof and Marangoni numbers is:

$$\Delta T = 0.1 \times \frac{q_{max}}{k} \, d \qquad (2.35b)$$

Regarding the convection mechanisms, the buoyancy effect within the melt and the thermocapillary effect on the free surface have been included in this model. The strength of buoyancy induced convection is indicated by the Rayleigh number, defined as

$$Ra = Gr \cdot Pr \qquad (2.36a)$$

where Gr is the Grashof number,

$$Gr = \frac{gr_b^3 \Delta \varrho}{\nu^2 \varrho} \qquad (2.36b)$$

Invoking the Boussinesq approximation, this becomes

$$Gr = \frac{g\beta_T \Delta T r_b^3}{\nu^2} \qquad (2.37)$$

where Pr is the Prandtl number, as defined in Eq. (2.26c). The relative magnitude of the thermocapillary effect is described by the Marangoni number, Eq. (2.28). The governing equations may be nondimensionalized with respect to the reference scales described at the start of this section as follows:

(i) continuity:

$$\frac{\partial(R\tilde{u})}{\partial R} + \frac{\partial(R\tilde{v})}{\partial Y} = 0 \qquad (2.38a)$$

(ii) radial momentum:

$$\frac{\partial(R\tilde{u}^2)}{\partial R} + \frac{\partial(R\tilde{u}\tilde{v})}{\partial Y} = -R\frac{\partial P}{\partial R} + Pr\left[\frac{\partial}{\partial R}\left(R\frac{\partial \tilde{u}}{\partial R}\right) + \frac{\partial}{\partial Y}\left(R\frac{\partial \tilde{u}}{\partial Y}\right)\right] - Pr\frac{2\tilde{u}}{R} \qquad (2.38b)$$

(iii) axial momentum:

$$\frac{\partial(R\tilde{u}\tilde{v})}{\partial R} + \frac{\partial(R\tilde{v}^2)}{\partial Y} = -R\frac{\partial P}{\partial Y} + Pr\left[\frac{\partial}{\partial R}\left(R\frac{\partial \tilde{v}}{\partial R}\right) + \frac{\partial}{\partial Y}\left(R\frac{\partial \tilde{v}}{\partial Y}\right)\right] + R \ Ra \ Pr \ \Theta \qquad (2.38c)$$

(iv) energy:

$$\frac{\partial(R\tilde{u}\Theta)}{\partial R} + \frac{\partial(R\tilde{v}\Theta)}{\partial Y} = \left[\frac{\partial}{\partial R}\left(R\frac{\partial \Theta}{\partial R}\right) + \frac{\partial}{\partial Y}\left(R\frac{\partial \Theta}{\partial Y}\right)\right] \qquad (2.38d)$$

where Ra is the Rayleigh number,

$$Ra = \frac{g\beta_T \Delta T r_b^3}{\nu\alpha} \qquad (2.38e)$$

and the italics denote the nondimensional variables. Here $\tilde{u}$ and $\tilde{v}$ are the velocity components in the radial and axial directions, respectively. The Boussinesq approximation is employed; thus, density variations are neglected in the governing equations, except for the buoyancy terms

47

in the momentum equations. Only the right half of the domain is considered in the calculation procedure, and symmetry boundary conditions are applied at the centerline. The associated boundary conditions are listed below in nondimensional form:

symmetry: $\quad\quad\quad\quad At\ R = 0,\ \tilde{u}\ =\ \dfrac{\partial \tilde{v}}{\partial R}\ =\ \dfrac{\partial \Theta}{\partial R} = 0$ $\quad\quad\quad$ (2.39a)

top and bottom boundaries:
$$At\ Y = 0 \text{ and } 1:\ \Theta = 0,\ \tilde{u} = \tilde{v} = 0 \quad\quad\quad (2.39b)$$

At the free surface; $R = f(Y)$:
(i) heat flux through the free surface (dimensionless) :
$$\vec{q} = \nabla\Theta \cdot \vec{n} \quad\quad\quad (2.39c)$$

which equals the normal derivative of the temperature at the free surface, and $\hat{n}$ is the unit normal vector to the free surface.

(ii) Marangoni effect:
$$\nabla\ (\vec{V} \cdot \vec{t}) \cdot \vec{n} = \text{Ma}\ \nabla\Theta \cdot \vec{t} \quad\quad\quad (2.34)$$

where the shear stress equals the surface tension gradient and gives rise to the normal derivative of the tangential velocity at the free surface. $\vec{V}$ is the velocity vector, $\vec{n}$ is the unit normal vector to the free surface, and $\vec{t}$ is the unit tangent vector to the free surface.

(iii) kinematic condition at the free surface:
$$\vec{V} \cdot \vec{n} = 0 \quad\quad\quad (2.39e)$$

which specifies a no-penetration condition at the free surface.

2.3.6   Results and Discussion

2.3.6.1   Prediction of Meniscus Shapes

In practice, the geometrical configurations employed result in a higher value of the Bond number. Consider the following configuration consisting of a feed rod diameter of 25 mm and a surface tension of 0.69 Nm^{-1} for Al_2O_3 and of 1.35 Nm^{-1} for NiAl (Keene 1993), resulting in Bond numbers of 7.6 and 5.7, respectively. Thus the Bond numbers lie in the range 5–10.

The meniscus shape is substantially influenced by the value of ΔP, the pressurization, in Eq. (2.18). In the literature, the value of the pressurization usually is determined by satisfying a constant volume criterion in addition to the boundary conditions imposed. In the present calculations, the objective was to find a range of pressurization values for which aspect ratios close to unity can be achieved. For a Bond number of 8.8, which was selected in order to effect

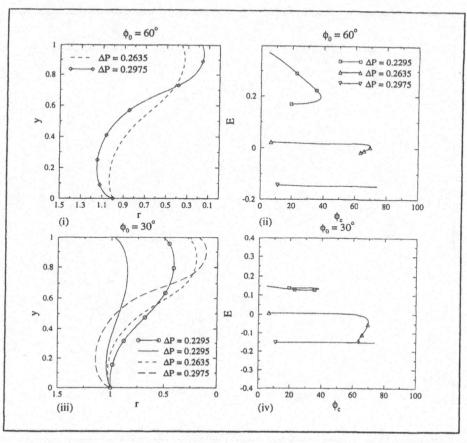

Figure 2.10 *Calculations for Bond number of 8.8, unity aspect ratio and imposed contact angle condition.*

(i) and (iii) Profiles meeting a desired contact condition of 60 deg and 30 deg respectively. (ii) and (iv) Free energy curves for imposed contact angles of 60 deg and 30 deg respectively, for the various values of the pressurization considered.

The curves in (ii) and (iv) use the same convention to indicate the pressurization value. (ii) and (iv) also indicate the range of values of ϕ_c that are obtainable by integrating the Young-Laplace equation. It may be noted that for $\phi_o = 30$ deg, and $\Delta P = 0.2295\%$, there are two profiles satisfying the given contact condition, indicated by the solid curves. The profile marked with $\circ$ does not minimize the free energy whereas the unmarked one does. None of the other profiles corresponds to a local minimum indicating that although they mathematically satisfy the Young-Laplace equation, they are not statically stable.

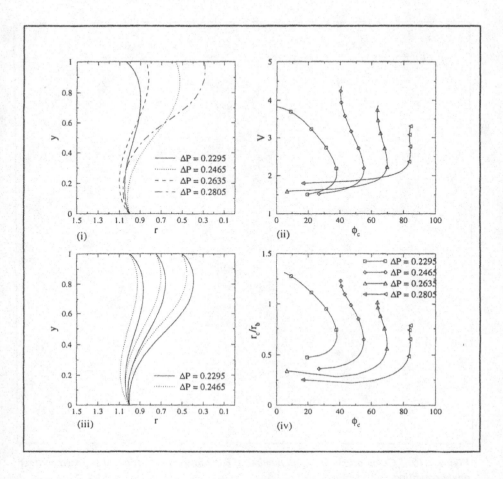

Figure 2.11   Calculations for Bond number of 8.8, unity aspect ratio and either (a) imposed radius at the top or (b) imposed volume of the melt.
Profiles for (i) fixed volume and (iii) fixed r_c (for three different r_c).
(ii) volume versus ϕ_c curves and
(iv) r_c/r_b versus ϕ_c.
The curves in (ii) and (iv) use the same convention to indicate the pressurization value. (ii) and (iv) also indicate the range of values of ϕ_c that are obtainable by integrating the Young–Laplace equation.

a comparison with the results of Quon et al. (1993), it was found that aspect ratios of unity could be achieved only in the range $0.2 \leq \Delta P \leq 0.3$. The pressurization ranges obtained here are used to generate the subsequent solutions.

Calculations were carried out for aspect ratio of unity, Bond number of 8.8, and six values of pressurization in the range $0.2 \leq \Delta P \leq 0.3$. Figures 2.10 and 2.11 show the results of the calculations.

Figure 2.10 shows the profiles and the corresponding free energy curves that are obtained when contact angles of 30 deg and 60 deg are imposed. The free energy curves also indicate the range of contact angles for which the Young-Laplace equation can be integrated. It is useful to draw attention to the case of $\phi_0 = 30$ deg, $\Delta P = 0.2295\%$. It may be observed that two profiles are obtained, marked in Fig. 2.10(iii) with the solid curves, one of them also marked with 'o'. The corresponding free energy curve in Fig. 2.10(iv) is marked with '□' and shows that two solutions can be obtained, one of which satisfies a local minimum. In Fig. 2.10(iii), the unmarked solid curve satisfies a local minimum indicating static stability. For $\phi_0 = 60$ deg, none of the obtained profiles satisfies a local minimum.

Figure 2.11 shows profiles that (i) meet a fixed volume constraint of $V = 2$ and 3 (nondimensionalized) and (iii) meet a fixed radius (r_c) condition for $r_c/r_b = 1.0, 0.75$, and 0.5. The corresponding r_c versus ϕ_c and V versus ϕ_c curves show the overall range over which the Young-Laplace equation can be integrated. It also may be noted that the overall solution behavior does not show any abrupt changes as the pressurization ΔP is varied over the specified range. Hence, the curves shown in Fig. 2.11(ii) help in determining the existence of solutions that can simultaneously satisfy a given contact angle condition and a fixed volume. The free energy curves, shown in Fig. 2.10(iv), determine the static stability of the calculated meniscus profiles.

2.3.6.2 Heat Transfer Calculations

A series of calculations was carried out with the geometry of the free surface being determined by the case with the parameters Bo = 8.8, $\Delta P = 0.2\%$, aspect ratio of unity and with $r_c = r_b$, as shown in Figure 2.11. The material selected for simulation is NiAl. The thermophysical properties of this material are listed in Table 2.1. If phase change effects are not considered, and the power input from the heater is 2.3×10^6 W/m², then the temperature scale can be obtained from Eq. (2.35b) as $\Delta T = 36.5$ K. For the length scale defined by $r_b = 1.25 \times 10^{-2}$ m as in the previous calculations for the meniscus shape, this temperature scale implies a Marangoni number of 10^3, a Grashof number of 520, and a Prandtl number of 0.04.

Table 2.1 Thermophysical properties of NiAl and process parameters

Density	6000 kg/m^3
Thermal conductivity	80 W/m-K
Heat capacity	586 J/kg-K
Coefficient of thermal expansion	1.51×10^{-5} K^{-1}
Kinematic viscosity	9×10^{-7} m^2/s
Surface tension coefficient	-2.7×10^{-4} N/m-K
Surface tension	1.35 N/m
Latent heat	5.1×10^5 J/kg
Melting point	1900 K
Gravitational acceleration	9.81 m/s^2

These values may vary with the composition of the NiAl. Nevertheless, it is clear that for the length scales under consideration, thermocapillary convection plays a dominant role in the heat transfer process. In the computations to follow, a range of nondimensional parameters,

$$0 \leq Gr \leq 10^6, \quad 0 \leq Ma \leq 10^3, \quad \text{and} \quad Pr = 0.04, \ 0.1$$

have been considered for investigation.

2.3.6.3 Numerical Procedure

The governing equations (2.38) are solved using the aforementioned pressure-correction formulation in nonorthogonal body-fitted coordinates. The convection and diffusion terms in both the momentum and energy equations are discretized using second order central differences. The grid points were clustered towards the free surface and for the computations involving Marangoni convection, they also were clustered towards the lower boundary at y = 0. For the cases involving natural convection only, two grid systems involving 81 x 52 nodes and 161 x 103 nodes, were used to discretize the domain. The solutions are indistinguishable, indicating grid independence. The solutions presented are based on the grid system involving 161 x 103 nodes for cases not involving Marangoni convection and 401 x 201 nodes for cases involving Marangoni convection.

2.3.6.4 Heat Conduction Only

A calculation was carried out with the heat transfer taking place through conduction effects only. The purpose was to carry out an a posteriori verification of the temperature scales used to define the Grashof and the Marangoni numbers in subsequent calculations. It may be noted that due to the curved geometry of the free surface, the location of the maximum temperature shifts

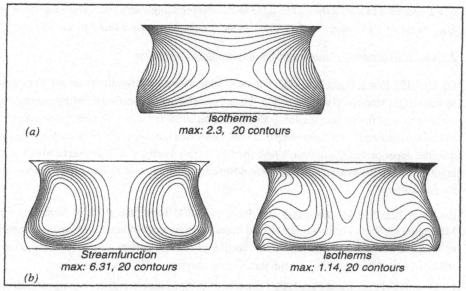

Figure 2.12. Effect of Natural convection on heat transfer in the float zone.
(a). Conduction case. The location of maximum temperature, Θ_{max}, indicates the effect of boundary curvature.
(b). Gr = 10^6 and Pr = 0.1. Buoyancy driven convection reduces the maximum temperature and shifts its location upwards, following the direction of the velocity field at the boundary. The notable feature is a single counterclockwise rotating convection roll whose spatial extent is of the order of the dimensions of the melt region.

towards the convex portion of the free surface, as is evident from Figure 2.12(a). The nondimensional value of Θ_{max} obtained here is 2.3 based on the temperature scale defined in Eq. (2.35b).

2.3.6.5 Natural Convection

Figure 2.12(b) shows the results of incorporating the effects of buoyancy induced convection with a Grashof number of 10^6 and a Prandtl number of 0.1. The streamline pattern shows a single recirculating zone in each half of the domain and is tangential to the free surface. It may be recalled that since surface tension effects have not been accounted for, the free surface does not exert a shear stress on the fluid. The fluid adjacent to the free surface is heated and rises to the top of the domain due to buoyancy effects, causing the isotherms to be distorted towards the top of the domain and the location of Θ_{max} to shift correspondingly upwards. The magnitude of Θ_{max} drops to 1.14, reflecting the validity of the temperature scale defined by Eq. (2.35b).

2.3.6.6 Interaction of Natural and Thermocapillary Convection

(a) $Gr = 10^6$, $Pr = 0.1$, and $Ma = 500$ Figure 2.13(a) shows the streamfunction and isotherms by imposing a Marangoni number of 500. Since the surface tension decreases with temperature, the tendency of the surface tension gradient induced shear stress is to reduce the convection strength in the vicinity of the convex part of the free surface and increase the convection strength near the upper part of the domain where the free surface is concave. The overall effect is to marginally shift the location of Θ_{max} in the downward direction, towards the convex portion of the free surface.

(b) $Gr = 10^6$, $Pr = 0.1$, and $Ma = 1000$ Figure 2.13(b) shows the effect of increasing the Marangoni number. The trends established in the previous case continue, shifting the location of Θ_{max} downwards towards the convex portion of the free surface. Since the convection strength in the vicinity of Θ_{max} is further reduced, the value of Θ_{max} also increases.

(c) $Gr = 2000$, $Pr = 0.1$, and $Ma = 1000$ Figure 2.13(c) shows the effect of decreasing the relative strength of buoyancy induced convection versus Marangoni convection. In this case, the convection pattern is dominated by the Marangoni effect. The convection pattern consists of both counterclockwise rotating convection rolls at the top right and the bottom left of the melt adjacent to the free surface and clockwise rotating convection rolls at the bottom right and the top left of the domain. The convection rolls adjacent to the convex portion of the free surface, which occur at the lower portion of the boundary, are substantially stronger because of the stronger temperature gradients along the convex portion of the free surface. The location of Θ_{max} is close to that of the pure conduction solution as compared to the high Gr cases. The effect of Marangoni convection is to shift the location of Θ_{max} to the convex portion of the boundary.

(d) $Gr = 2000$, $Pr = 0.04$, and $Ma = 1000$ Figure 2.14(d) shows the effect of decreasing the Prandtl number. The effect of lower Prandtl number is to increase the convective effects in the flow (Shyy, 1994) by decreasing the magnitude of the viscous terms in the momentum equations. In the energy equation, this effect will be balanced by the increased thermal diffusivity. The convection pattern is qualitatively similar to the case shown in Fig. 2.13(c). It can be observed that the isotherms are distorted near the convex portion of the free surface, due to the strong fluid convection, whereas in the upper portion of the domain, the pattern resembles the pure conduction case.

From the above cases, some useful inferences can be made that are directly applicable to the float-zone growth of NiAl. For the length scales of current experimental interest, it is

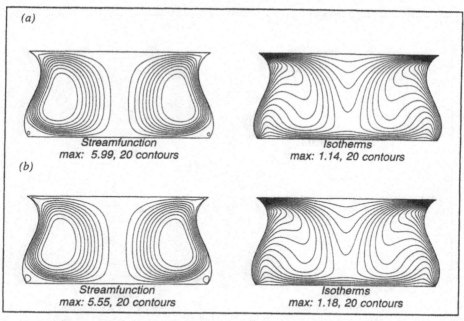

Figure 2.13. *Effect of Rayleigh number, Marangoni number and Prandtl number on transport characteristics in the float zone.*

(a). $Gr = 10^6$, $Ma = 500$ and $Pr = 0.1$. The onset of surface tension driven convection marginally decreases the overall convection strength and marginally shifts the location of Θ_{max} in the negative z direction. The tendency of the surface tension gradient is to induce a clockwise convection roll in the lower portion of the domain and a counterclockwise roll in the upper portion of the domain, the spatial extent of which is confined to the region adjacent to the free boundary compared with the case of buoyancy driven convection.

(b). $Gr = 10^6$, $Ma = 1000$ and $Pr = 0.1$. Strong surface tension effects cause the overall convection strength to decrease and Θ_{max} to increase and shift further in the downward direction. The tendency of the surface tension gradient is to induce a clockwise convection roll in the lower portion of the domain and a counterclockwise roll in the upper portion of the domain.

observed that the pure conduction model of heat transfer is inadequate. Strong convective heat transfer effects can be observed, which are dominated by the Marangoni effect. The Marangoni effect causes clockwise recirculating convection rolls in the bottom right and top left of the domain and counterclockwise recirculating rolls at the top right and bottom left of the domain. For the high Bond numbers that prevail under 1-g conditions on earth, the free surface has a convex shape in the lower portion of the melt due to the hydrostatic pressure of the melt column. This convex shape induces asymmetry in the shape, spatial extent, and strength of the convection

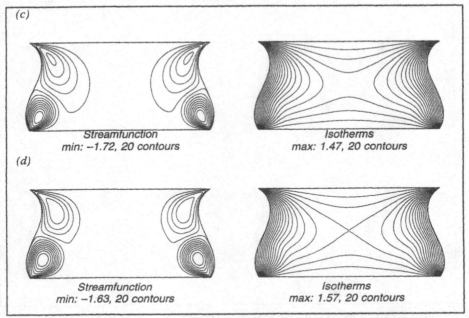

Figure 2.13. (continued). Effect of Rayleigh number, Marangoni number and Prandtl number on transport characteristics in the float zone.

(c). *Gr = 2000, Ma = 1000 and Pr = 0.1. With decreasing Grashof number, the lower clockwise rotating convection cell due to the Marangoni effect becomes more evident. The location of the maximum temperature shifts downward because of the convex shape of the lower part of the free surface. As a result, the convection in the upper and central regions is substantially weaker than the cases dominated by buoyancy induced convection.*

(d) *Gr = 2000, Ma = 1000 and Pr = 0.04. With decreasing Prandtl number, the magnitude of the maximum temperature increases, but the convection cells penetrate less into the melt zone and the overall convection strength is reduced.*

rolls and the distribution of the temperature. For the pure conduction case, the convex shape causes the location of the maximum temperature to be shifted towards the convex side of the free surface. In the presence of Marangoni convection, this effect is augmented by the strong convective effect near the convex portion of the domain, due to the stronger temperature gradients along the convex portion of the free surface. This convection in turn causes the location of the maximum temperature to move further downwards. Marangoni convection completely overwhelms the buoyancy-driven convection for this particular configuration and material properties, augmenting the buoyancy effect in the upper portion of the domain and counteracting it in the lower portion of the domain. Thus, the dependence of surface tension on temperature plays a dominant role in the heat transfer characteristics in the float zone.

2.3.7 Effect of Convection on Meniscus Shape

The deformation of the free surface due to thermocapillary convection is dominated by the Capillary number. For example, Zebib et al. (1985) found that whereas the shape is qualitatively sensitive to the Prandtl number, the magnitude of the deformation is determined, to the leading order, by the Capillary number. More insight into the effects of Marangoni convection on transport dynamics can be found in Carpenter and Homsy (1989) and Mundrane and Zebib (1994). Equation (2.24) describes the role of the normal stress, arising from the normal derivative of the normal velocity component at the free surface, in changing the shape of the meniscus. Under 1-g conditions and at the length scales considered, the hydrostatic terms dictated by the high Bond numbers are the dominant terms in Eq. (2.24) and the normal stress terms due to convection are small compared to the hydrostatic term. The Bond number for this case is Bo = 8.8 and the Capillary number is Ca = 1.5 × 10^{-5}. The impact on the surface curvature varies with the vertical distance and is described by $1 + \dfrac{Ca}{Bo} \dfrac{\partial V_n/\partial n}{Y}$. This quantity may be plotted to show the variation of its magnitude along the y-axis as shown in Figure 2.14. This shows that except at the lowermost corners of the domain, the normal stress term is vanishingly small. Considering that the meniscus is pinned at both the top and the bottom of the domain, the correction to the overall shape of the meniscus is expected to be modest except at the lowermost corner where the curvature will be substantially lower. Since the effect is extremely localized at the corner locations where the flow turns to match the boundary, the overall effect on the heat transfer and convection characteristics is expected to be negligible for this case. Accordingly, corrections to the meniscus shape appear unnecessary in this particular study. We also observe that the percentage variation of the surface tension with temperature is less than 1% and hence cannot significantly affect the force balance at the interface.

Similar results have been reported by Tao et al. (1995), where a vorticity-stream function formulation was employed. The fluid chosen in that study is silicone oil, which has a very different Prandtl number, of the order of 10^2 to 10^4. Their solutions can help understand Marangoni convection in a different parameter range. In a recent effort, Hou et al. (1994) have employed a boundary integral approach to simulate surface tension driven flows, with good results obtained for a wide range of phenomena. Kanouff and Greif (1994) have analyzed oscillations in thermocapillary convection in a square cavity.

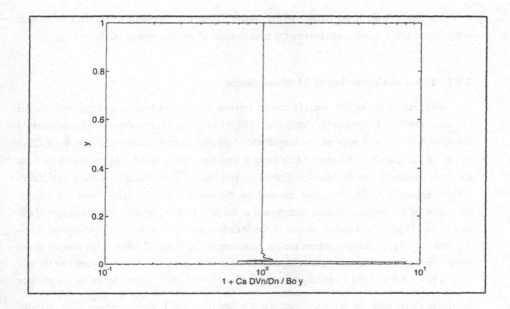

Figure 2.14 Semilog plot of Y versus $1 + \dfrac{Ca}{Bo}\dfrac{\partial V_n/\partial n}{Y}$ *to show the locations where the normal stress term due to convection becomes important compared to the hydrostatic pressure.*

2.4 CONCLUSIONS

In this chapter, the salient features of a moving grid formulation for the solution of the fluid flow equations were presented. Although the pressure-based algorithm was our focus here, other techniques such as those mentioned at the beginning of this chapter could be employed as well. It was emphasized that regardless of the particular fluid flow solution algorithm to be utilized, care must be exercised both in calculating the geometric information in handling the conversion of the velocity components from Cartesian to curvilinear coordinates and in defining the free boundary shapes and locations. These numerical issues must be addressed before one can consider those related to the physical aspects of moving boundaries.

In the following chapters, we shall discuss additional issues dealing with the incorporation of moving boundaries into the fluid flow solution procedure; again, the pressure-based algorithm will be used to facilitate our presentation and illustration. The moving boundary techniques to be discussed include fixed grid, moving grid, and combined

Eulerian-Lagrangian technique to track moving interfaces. Various measures and algorithmic details involved in performing such simulations are presented.

Test cases with analytical solutions available as well as practical applications involving complex physical mechanisms will be used to assess the performance and relative merits of each technique and to help highlight the modeling and computational issues involved. Solutions for physical problems encompassing wide length scale variations will be presented. Using case studies, we intend to present a complete picture of these computational techniques, including theoretical formulation of the physical problem, nondimensionalization and scaling, coupling of moving boundaries and fluid dynamics, and physical interpretation. The examples chosen are mainly from materials processing and fluid flow–flexible structure interaction. Although physical examples can be drawn from many other areas, they will not be presented in order to maintain the focus of the presentation.

MOVING GRID TECHNIQUES: FLUID MEMBRANE INTERACTION

3.1   DESCRIPTION OF THE PHYSICAL PROBLEM

The interaction of fluid flows with solid structures is a problem of considerable interest from a technological standpoint (Löhner et al. 1995). In the area of aerodynamics, for example, flutter problems have been actively researched for several decades now (Bisplinghoff et al. 1955, Rizzetta 1979, Dowell 1980, Edwards et al. 1983, Bartina 1989, Lu et al. 1995). On the other hand, thin–film coating flows over a deformable substrate, such as paper, photographic film, extensible membrane, or magnetic tape, represent the same type of physical problem (Marchaj 1979, Ruschak 1985, Christodoulou and Scriven 1989, Probstein 1989, Aidun 1991). These are challenging problems because the time and length scales which govern the behavior of fluids and solids can be quite different, and the movement of solid structures generally causes computational difficulties in determining the flow and stress fields and their coupling. In particular, if the solid structure is not fixed or is deformable, or if the impacting fluid flow induces motion of the solid structure, the analysis of such interactions becomes complicated.

One interesting situation in which fluid flow interacts strongly with a non–stationary, deformable solid structure occurs in the operation of membrane wings or sails. Hitherto, the design and performance of sails have benefitted little from detailed analysis of the fluid/membrane interaction and have relied largely on hands–on experience and practical demonstration. Marchaj, in his second book on the science of sailing (Marchaj 1979), begins the section on sail design with the following comments.

Despite the fact that mathematics, computers, and wind tunnel testing are playing an increasing part in the designing of sails, sailmaking as well as sail tuning are still strongholds of art based on a hit-or-miss technique rather than on science.... After all, unlike the aeroplane wing, which can be regarded as a rigid structure whose shape is unaffected by variation in incidence and speed, sail shape is a function of both, in which the shape of the sail affects the pressure distribution and vice versa, in a rather unpredictable manner.

These comments by Marchaj suggest two things concerning the analysis of marine sails. First, the behavior and performance of a sail is governed by the aeroelastic interaction of a fluid dynamic field and a deformable surface. Second, as a result of this interaction, as well as the presence of other complexities, analytic and computational methods which have enjoyed considerable success in the design of aircraft have not yet proven useful to sail designers. It is likely that the second observation concerning the usefulness of analytic methods to sail designers will not always hold true.

As an illustration of the potential usefulness of computational methods in the analysis of sails, consider the membrane wing of Fig. 3.1, which is shown operating near a boundary such as the surface of the sea. Since the sail is located very near the boundary it is immersed in the shear layer adjacent to the free surface. The kinematic inflow conditions to the sail consequently are nonuniform and also may be unsteady due to the presence of wind gusts. Furthermore, since the sail is not stationary but rather has some forward velocity, the relative inflow velocity far upstream of the sail will vary in both magnitude and direction along the sail span, as well as varying in time. Computational methods provide a means for simulating such a complex flow environment which would otherwise be nearly impossible to reproduce experimentally.

Having introduced the problem of marine sail aerodynamics, the moving grid technique presented in Chapter 2 is now applied to the simulation of viscous, laminar flow over a flexible membrane wing. The present and next chapters treat two different classes of physical problems with the same computational approach, namely, the moving grid technique, to demonstrate the common features of moving boundary problems in different physical contexts. Here, the coupling of the flow field to the movement and deformation of a flexible membrane offers a stringent test of the robustness, efficiency and accuracy of the computational algorithm because the fluid and solid domains contain different physical characteristics and time scales; the movement of the membrane wing also poses serious computational challenges. For this problem both steady-state and time-accurate computations are presented. In the steady-state case, the moving grid procedure is still employed–albeit not in a time-accurate sense–since the initial membrane configuration and associated body-fitted grid do not correspond to an equilibrium

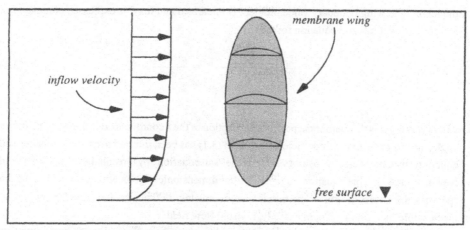

Figure 3.1 Schematic of a membrane wing of finite span operating near a free surface.

configuration. Consequently, the grid is continuously updated during the course of the aeroelastic solution in response to adjustments in the membrane wing configuration. In the time-dependent case, an equilibrium membrane configuration and a corresponding grid is determined at each time step so that both the unsteady fluid dynamic conservation laws and membrane equilibrium equations are satisfied. Before the details of the membrane wing problem are discussed, the relevant background information is presented.

3.1.1 Potential Flow–Based Membrane Wing Models

The vast majority of the published works related to membrane wing aerodynamics have made several simplifying approximations concerning both the elastic characteristics of the membrane itself as well as the nature of the surrounding flow field. Perhaps the most significant of these simplifying assumptions is that the fluid dynamics can be adequately described by a potential-based description of the flow field. In addition to the almost universally adopted potential flow assumption, the additional approximations associated with thin airfoil theory–small camber and incidence angle–also are often made and the membrane itself generally is considered to be inextensible.

The analysis of membrane wings begins with the historical works of Voelz (1950), Thwaites (1961), and Nielsen (1963). These works consider the steady, two-dimensional, irrotational flow over an inextensible membrane with slack. As a consequence of the inextensible assumption and the additional assumptions of small camber and incidence angle,

63

the membrane wing boundary value problem is linearized and may be expressed compactly in nondimensional integral equation form as

$$1 - \frac{C_T}{2} \int_0^1 \frac{\frac{d^2(y/\alpha)}{d\zeta^2}}{2\pi(\zeta - x)} d\zeta = \frac{d(y/\alpha)}{dx} \tag{3.1}$$

where $y(x)$ defines the membrane profile as a function of the x coordinate, α is the flow incidence angle, and C_T is the tension coefficient. Equation (3.1) has been referred to as the "Thwaites sail equation" by Chambers (1966) and simply as the "sail equation" by Greenhalgh et al. (1984) and Newman (1987). This equation, together with a dimensionless geometric parameter ε which specifies the excess length of the membrane, completely defines the linearized theory of an inextensible membrane wing in a steady, inviscid flow field.

Different analytical and numerical procedures have been applied to the basic equation set in order to determine the membrane shape, aerodynamic properties, and membrane tension in terms of the angle of attack and excess length. In particular, Thwaites (1961) obtained eigensolutions of the sail equation which are associated with the wing at an ideal angle of incidence. Nielsen (1963) obtained solutions to the same equation using a Fourier series approach which is valid for wings at angles of incidence other than the ideal angle. Other more recent but similar works are those by Greenhalgh et al. (1984), Sugimoto and Sato (1988), and Vanden-Broeck and Keller (1981).

Various extensions of the linear theory have appeared in the literature over the years. Vanden-Broeck (1982) and Murai and Maruyama (1980) developed nonlinear theories valid for large camber and incidence angle. The effect of elasticity has been included in the membrane wing theories of Jackson (1983) and Sneyd (1984) and the effects of membrane porosity have been investigated by Murata and Tanaka (1989). In a paper by de Matteis and de Socoi (1986), experimentally determined separation points were used to modify the lifting potential flow problem in an attempt to model flow separation near the trailing edge. The effects of elasticity and porosity also were considered in this work. A comprehensive review of the work published prior to 1987 related to membrane wing aerodynamics is given by Newman (1987).

The agreement between the various potential flow-based membrane wing theories and experimental data has been reported by several authors including Greenhalgh et al. (1984), Sugimoto and Sato (1988), and Newman and Low (1984). In general, there has been considerable discrepancy between the measurements made by the different authors (Jackson 1983), which have all been in the turbulent flow regime at Reynolds numbers between 10^5 and 10^6. As a result of the discrepancies in the reported data—due primarily to differences in Reynolds number and experimental procedure—the agreement between the potential based membrane theories and the data has been mixed. In particular, the measured lift is in fair agreement with the predicted value when the excess length ratio is less than 0.01 and the angle of attack is less than 5°. However, even for this restricted range of values, the measured tension is significantly less than predicted by theory. Furthermore, for larger excess lengths and incidence angles the lift and tension are poorly predicted by the theory.

Furthermore, for larger excess lengths and incidence angles the lift and tension are poorly predicted by the theory. Flow visualization studies indicate that the main reason for the disagreement is the existence of a thick boundary layer or region of separated flow on the membrane, typically near the trailing edge. It has been noted by several authors that the presence of viscous effects such as thick boundary layers and separation regions will overshadow any implications associated with the linearizing approximations made by thin wing theory (Newman 1987). Recently, progress has been made to employ the full Navier–Stokes equations for this problem in order to take the viscous effect into account (Smith and Shyy 1995a, b); this approach will be presented in the following.

3.1.2 Membrane Equilibrium

In this section the general equilibrium equations are presented for a two-dimensional elastic membrane subjected to both normal and shearing stresses. The membrane is assumed to be massless, and the equilibrium conditions are stated in terms of the instantaneous spatial Cartesian coordinates and the body-fitted curvilinear coordinates. The basic formulation is essentially identical to many previously published works such as de Matteis and de Socoi (1986) and Sneyd (1984).

Figure 3.2 illustrates an elastic membrane restrained at the leading and trailing edges and subjected to both normal and tangential surface tractions p and τ, respectively. Imposing equilibrium in the normal and tangential directions requires

$$\frac{d^2y}{dx^2}\left[1 + \left(\frac{dy}{dx}\right)^2\right]^{-\frac{3}{2}} = -\frac{\Delta p}{\gamma} \tag{3.2a}$$

$$\frac{dy}{d\xi} = -\tau \tag{3.2b}$$

where γ is the membrane tension. Equation (3.2a) is the Young-Laplace equation cast in Cartesian coordinates. The corresponding form in cylindrical coordinates is given by Eq. (2.18). The net pressure and shear stress acting on a segment of the membrane are given respectively by

65

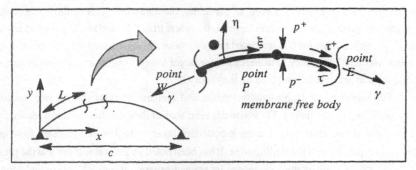

Figure 3.2 End constrained elastic membrane

$$\tau = \tau^- + \tau^+ \tag{3.3b}$$

where the superscripts indicate the values at the upper and lower surfaces of the membrane, as shown in the figure. If the membrane material is assumed to be linearly elastic, the nominal membrane tension $\bar{\gamma}$ may be written in terms of the nominal membrane strain $\bar{\delta}$ as

$$\bar{\gamma} = \left(S^0 + E\bar{\delta} \right)h \tag{3.4}$$

where S^o is the membrane prestress, E is the elastic modulus, and h is the membrane thickness. The nominal membrane strain is given by

$$\bar{\delta} = \frac{L - L_0}{L_0} \tag{3.5}$$

where L_0 is the unstrained length of the membrane and L is the length of the membrane after deformation, which may be expressed in terms of the spatial Cartesian coordinates as

$$L = \int_0^c \sqrt{1 + \left(\frac{dy}{dx}\right)^2}\, dx \tag{3.6}$$

where c is the chord length. At the leading and trailing edges of the membrane, the following boundary conditions are imposed on Eq. (3.2a)

$$y = 0 \quad \text{at} \quad x = 0, c \tag{3.7}$$

A discrete form of the elastic membrane boundary value problem can be obtained at a finite number of points on the fixed interval $[0,c]$ by replacing the derivatives in Eq. (3.2) and the integral in Eq. (3.6) with appropriate finite difference and finite sum approximations. Applying central difference approximations to Eq. (3.2a) leads to a three-point difference stencil centered around point P with neighboring points E and W, as shown in Figure 3.2. At each point P, an equation of the general form

$$a_P y_P = a_E y_E + a_W y_W + S_y \qquad (3.8)$$

then is obtained where the coefficients a's are associated with the finite difference approximation, and S_y is a source term containing all terms in Eq. (3.2a) that cannot be expressed as a linear combination of the Cartesian coordinates. Consequently, all of the nonlinearity in the boundary value problem is contained in the source term. The resulting set of finite difference equations then may be solved using a line iteration method with underrelaxation. As a measure of the degree to which the discrete form of Eq. (3.2a) has been satisfied, the residual of the membrane equilibrium equation is defined as follows:

$$R_y = \sum_{all\ P} \left| - a_P y_P + a_E y_E + a_W y_W + S_y \right| c \qquad (3.9)$$

3.1.3 Nondimensionalization of the Governing Equations

The aeroelastic boundary value problem can be written in nondimensional form after introducing the following dimensionless variables:

$$\bar{u} = \frac{u}{V_\infty} \qquad (3.10a)$$

$$\bar{v} = \frac{v}{V_\infty} \qquad (3.10b)$$

$$\bar{t} = t\omega \qquad (3.10c)$$

$$X = \frac{x}{c} \qquad (3.10d)$$

$$Y = \frac{y}{c} \qquad (3.10e)$$

$$P = \frac{p}{\varrho V_\infty^2} \qquad (3.10f)$$

$$\Delta P = \frac{\Delta p}{q_\infty} \qquad (3.10g)$$

$$\hat{\gamma} = \frac{\gamma}{S^0 h} \qquad (3.10h)$$

or

$$\hat{\gamma} = \frac{\gamma}{Eh} \qquad (3.10i)$$

where either Eq. (3.10h) or Eq. (3.10i) is used to nondimensionalize the membrane tension depending on whether the tension is dominated by pretension or by elastic strain. In Eq. (3.10), V_∞ is the freestream velocity, q_∞ is the freestream stagnation pressure equal to $\frac{1}{2}\varrho V_\infty^2$, ω is the freestream frequency, and c is the chord length. Substituting these variables into Eq. (3.2a) leads to the following dimensionless equilibrium equation when membrane tension is dominated by elastic strain:

$$\frac{d^2Y}{dX^2}\left(1 + \left(\frac{dY}{dX}\right)^2\right)^{-\frac{3}{2}} = -\left(\frac{1}{\Pi_1}\right)^3 \frac{\Delta P}{\hat{\gamma}} \tag{3.11a}$$

with Π_1 defined to be
$$\Pi_1 = \left(\frac{Eh}{q_\infty c}\right)^{\frac{1}{3}} \tag{3.11b}$$

When membrane tension is dominated by pretension, Eq. (3.2a) leads to

$$\frac{d^2Y}{dX^2}\left(1 + \left(\frac{dY}{dX}\right)^2\right)^{-\frac{3}{2}} = -\left(\frac{1}{\Pi_2}\right)\frac{\Delta P}{\hat{\gamma}} \tag{3.11c}$$

with Π_2 defined to be

$$\Pi_2 = \left(\frac{S^0 h}{q_\infty c}\right) \tag{3.11d}$$

The boundary conditions also may be written in nondimensional form as
$$Y = 0 \text{ at } X = 0, 1 \tag{3.12}$$

Regarding the physical significance of the aeroelastic parameters Π_1 and Π_2, we note that the nondimensional deformation of an initially flat elastic membrane is inversely proportional to Π_1 in the absence of pretension. Alternately, the deformation of a membrane is inversely proportional to Π_2 in the presence of large initial pretension. Consequently, the steady-state, inviscid aeroelastic response of an initially flat membrane wing at a specified angle of attack is controlled exclusively by Π_1 in the limit of vanishing pretension and exclusively by Π_2 in the limit of vanishing material stiffness.

Substituting the same dimensionless variables into Eq. (2.1) leads to the following nondimensional form of the incompressible Navier-Stokes equations

$$\frac{\partial(\rho\bar{u})}{\partial X} + \frac{\partial(\rho\bar{v})}{\partial Y} = 0 \tag{3.13a}$$

$$S_r \frac{\partial\bar{u}}{\partial t} + \frac{\partial(\bar{u}\bar{u})}{\partial X} + \frac{\partial(\bar{u}\bar{v})}{\partial Y} = -\frac{\partial P}{\partial X} + \frac{1}{Re}\left[\frac{\partial}{\partial X}\left(\frac{\partial\bar{u}}{\partial X}\right) + \frac{\partial}{\partial Y}\left(\frac{\partial\bar{u}}{\partial Y}\right)\right] \tag{3.13b}$$

$$S_r \frac{\partial\bar{v}}{\partial t} + \frac{\partial(\bar{u}\bar{v})}{\partial X} + \frac{\partial(\bar{v}\bar{v})}{\partial Y} = -\frac{\partial P}{\partial Y} + \frac{1}{Re}\left[\frac{\partial}{\partial X}\left(\frac{\partial\bar{v}}{\partial X}\right) + \frac{\partial}{\partial Y}\left(\frac{\partial\bar{v}}{\partial Y}\right)\right] \tag{3.13c}$$

with boundary conditions at the membrane surface

$$\tilde{u} = S_r \, \dot{X} \tag{3.14a}$$

$$\tilde{v} = S_r \, \dot{Y} \tag{3.14b}$$

and far upstream from the membrane

$$\tilde{u} = \cos\alpha \, (1 + \beta \sin \tilde{t}) \tag{3.15a}$$

$$\tilde{v} = \sin\alpha \, (1 + \beta \sin \tilde{t}) \tag{3.15b}$$

where β is a constant defining the magnitude of the oscillation about the mean value. The dimensionless parameters appearing in Eq. (3.13), the Reynolds number and the Strouhal number, are defined as

$$\text{Re} = \frac{V_\infty \varrho c}{\mu} \tag{3.16a}$$

$$S_r = \frac{\omega c}{V_\infty} \tag{3.16b}$$

When the membrane is not initially flat and taut, the geometry of the wing may be characterized by an additional dimensionless quantity, the excess length parameter ε, which is defined as

$$\varepsilon = \frac{L_0 - c}{c} \tag{3.17}$$

where L_0 is the unstrained length of the membrane. The set of dimensionless parameters given above–Π_1, Π_2, S_r, Re, ε, and α–completely characterize the physical problem considered in the present work.

In addition to the basic dimensionless parameter set, other physically significant nondimensional parameters may be derived. In particular, a frequency ratio parameter, Ω, may be introduced by drawing an analogy between a one-degree-of-freedom spring/mass system and the transverse motion of the membrane in response to a harmonically varying freestream velocity. If the nondimensional frequency parameter, is chosen as the ratio of system forcing frequency ω to the system natural frequency ω_n, that is,

$$\Omega = \frac{\omega}{\omega_n} \tag{3.18}$$

then the following parameters may be derived from the basic set and substituted for the Strouhal number in the basic parameter set:

69

$$\Omega_1 = \frac{S_r}{\Pi_1^{3/2}} \tag{3.19a}$$

$$\Omega_2 = \frac{S_r}{\Pi_2^{1/2}} \tag{3.19b}$$

Again, as in the case of steady flow, Ω_2 is the appropriate dimensionless frequency when the membrane is substantially pretensioned, whereas Ω_1 is the appropriate dimensionless frequency when the membrane develops tension elastically.

Finally, the aerodynamic lift, drag, and moment about the quarter chord and the membrane tension and fluid dynamic pressure are nondimensionalized in the customary way as stated below:

$$C_L = \frac{\mathcal{L}}{\frac{1}{2}\varrho V_\infty^2 c} \tag{3.20a}$$

$$C_D = \frac{\mathcal{D}}{\frac{1}{2}\varrho V_\infty^2 c} \tag{3.20b}$$

$$C_M = \frac{\mathcal{M}_{c/4}}{\frac{1}{2}\varrho V_\infty^2 c^2} \tag{3.20c}$$

$$C_\gamma = \frac{\bar{\gamma}}{\frac{1}{2}\varrho V_\infty^2 c} \tag{3.20d}$$

$$C_p = \frac{p - p_\infty}{\frac{1}{2}\varrho V_\infty^2} \tag{3.20e}$$

In Eqs. (3.20), the lift force $\mathcal{L}$, drag force $\mathcal{D}$, and quarter-chord moment $\mathcal{M}_{c/4}$ are obtained by integration of the aerodynamic pressure and shear stress over the membrane chord and resolving the net force vector into components parallel and perpendicular to the freestream velocity (McCormick 1979).

3.1.4   The Moving Grid Computational Procedure

Our primary objective is to determine the equilibrium configuration and associated aerodynamic characteristics of an elastic membrane wing as a function of time. Consequently, the present aeroelastic problem consists of finding membrane configurations and aerodynamic surface pressures and shear stresses which simultaneously satisfy Eq. (3.2) and Eq. (2.1), respectively. An iterative procedure is used to solve the coupled boundary value problems by computing the elastic and aerodynamic problems cyclically until a solution is obtained at each time step that satisfies the governing equations to a predetermined convergence criterion.

Since a fully implicit time-integration scheme is used to advance the solution from one time level to the next, all grid-dependent terms appearing in Eq. (2.2) must be reevaluated each time the membrane profile and field grid points are adjusted during the aeroelastic iteration procedure. The Cartesian components of the grid velocity also must be reevaluated since the grid velocity enters the solution not only through the definition of the contravariant velocity components but also through the kinematic boundary condition given in Eq. (3.14). Consequently, the metrics of the transformation as well as the components of the grid velocity vector are recomputed directly from the updated grid coordinates each time the membrane equilibrium equation is solved. An exception to this strategy is made when the Jacobian of the transformation at the implicit time level appearing in Eq. (2.5) is evaluated. This grid-dependent quantity is not computed directly from the updated grid coordinates but rather by using Eq. (2.17).

During the course of the aeroelastic iteration procedure, it is necessary to update the grid in the physical domain so that it remains body-fitted as the wing deforms. The maintenance of the body-fitted grid is achieved by updating the grid points in the vicinity of the membrane during the course of iteration according to

$$Y^{(i)} = Y^{(i-1)} + g(\eta)(Y^{(i)} - Y^{(i-1)})$$ (3.21a)

where $g(\eta)$, with η defining the grid index along the transverse coordinate, is a general function that decays with distance away from the membrane surface. The superscripts appearing in Eq. (3.21a) indicate the level of the aeroelastic iteration procedure. Presently, $g(\eta)$ is chosen as an exponential function defined as

$$g(\eta) = \exp\left(\frac{-|\eta|}{c_1}\right)$$ (3.21b)

where c_1 is a constant that depends on the grid resolution and the Reynolds number. A similar strategy using an exponentially decaying function for updating field grid points while maintaining a body-fitted curvilinear grid has been used by Boschitsch and Quackenbush (1993).

In Section 2, we apply the moving grid technique to investigate the interaction of a viscous flow and a membrane wing. Both steady-state and time-dependent problems will be considered. The membrane shape is determined by a balance of the fluid dynamic forces and the elastic forces developed within the membrane. Hence, the membrane shape is determined as a part of the solution. The moving grid technique is initiated by assigning an initial guess to the membrane shape and solving the fluid flow equations. A new membrane shape is obtained from the force balance, and the domain is remeshed and the fluid flow equations are solved again. This procedure is integrated into the solution sequence and repeated until convergence. This iterative procedure, which involves computing the domain shape, remeshing the domain, and recomputing the fluid flow equations, constitutes the moving grid technique.

3.1.5 A Potential Flow Model for Thin Wings

Before moving on to the topic of viscous flow interacting with a membrane wing, which is our main interest in this chapter, we present a method for computing the surface pressure distribution resulting from the irrotational, incompressible flow over thin, two-dimensional wings of arbitrary shape. The reason for including in the present work a numerical procedure for the solution of lifting potential flows is to highlight the difference in aerodynamic quantities predicted by a viscous flow based membrane wing model when compared to previous work on membrane wing aerodynamics based on a potential description of the flow field.

The fundamental assumption concerning the flow field around the wing is that it is irrotational, and consequently the velocity field may be derived from a scalar potential. Implicit in this assumption is that the flow is inviscid and free of vorticity far upstream. The use of point vortex singularities to model the lifting potential flow around thin wings has its historical origin and justification in work by James (1972). A description of the method in its modern form may be found in Katz and Plotkin (1991).

Figure 3.3 shows a membrane airfoil that has been discretized into a number of linear segments or panels. A typical segment is composed of a point vortex and a control point where the flow tangency condition is enforced. If the point vortices and control points are located at the local quarter and three-quarter chord points of the panel, respectively, the Kutta condition is implicitly satisfied and no additional boundary condition is needed at the trailing edge (James 1972, Katz and Plotkin 1991). By modeling the wing as an assemblage of vortex singularity segments of unknown strength and enforcing the zero normal relative velocity condition at each control point, designated as point i, the following set of linear algebraic equations may be formed and solved for the vortex strength Γ at point j.

$$A_{ij}\,\Gamma_j = (-\,\boldsymbol{V}_\infty \cdot \boldsymbol{n}\,)_i \qquad (3.22)$$

In Eq. (3.22), A_{ij} is a matrix of vortex influence coefficients, defined as the normal velocity induced at control point i by a unit strength vortex at point j, $\boldsymbol{V}_\infty$ is the freestream velocity vector, and $\boldsymbol{n}$ is the unit normal vector at the control point.

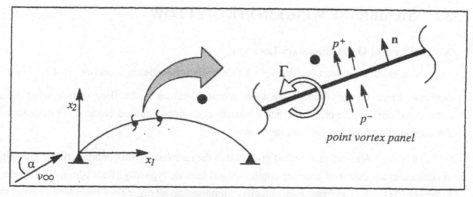

Figure 3.3. Potential flow wing model.

Once the vortex strengths have been determined from Eq. (3.22) the perturbation velocities on the upper and lower surfaces, respectively v'^+ and v'^-, are given by

$$v'^+ = \frac{\Gamma}{2\,l}\begin{bmatrix} n_y \\ -n_x \end{bmatrix} \tag{3.23a}$$

$$v'^- = -\frac{\Gamma}{2\,l}\begin{bmatrix} n_y \\ -n_x \end{bmatrix} \tag{3.24b}$$

where l is the length of the segment or panel and n_x and n_y are the Cartesian components of the unit normal vector.

With the velocity field determined on the upper and lower surfaces of the wing, the pressure on the wing surfaces, p^+ and p^-, may be computed directly from the Bernoulli equation as follows

$$p^{\pm} = \frac{1}{2}\varrho\,(\,V_{\infty}^2 - v^{\pm^2}\,) \tag{3.24}$$

Here, the fluid velocity on the upper and lower wing surfaces is the sum of the perturbation velocity and the freestream velocity and is given by

$$v^{\pm} = V_{\infty} + v'^{\pm} \tag{3.25}$$

It should be pointed out that the leading edge suction force arising from the pressure singularity at the leading edge of the infinitely thin membrane is not included in the above potential flow model.

3.2   MEMBRANE WINGS IN STEADY FLOW

3.2.1   Effect of Outer Boundary Location

We first present solutions obtained with a steady-state freestream condition, that is, V_∞ is a constant. Since the physical problem under consideration is the flow over a wing in an unbounded domain, the placement and boundary conditions imposed on the outer boundary of the computational domain require investigation.

Figure 3.4(a) shows a typical grid used in the membrane wing computations. The wing is situated in the center of a square computational domain typically $10c \times 10c$ in size, where c is the chord length. The freestream velocity is imposed on all outer flow boundaries except the downstream boundary, where a zero-gradient condition is imposed on the velocity components. An enlarged view of the grid in the vicinity of the wing is shown in Fig. 3.4(b). The refinement of the grid near the membrane was based primarily on experience and the degree to which flow separation was expected in each situation. Attention to grid refinement is particularly important near the leading and trailing edges and in the wake.

A series of computations was performed using different conditions at the outer domain boundary prior to selecting the boundary condition set shown in Fig. 3.4. These included computations using Dirichlet conditions at all outer domain boundaries as well as computations where a zero-gradient condition was imposed on all boundaries except the upstream boundary, where freestream conditions were imposed. In the latter case, with exit conditions specified at the north, south, and east domain boundaries, the algorithm was unable to admit a uniform flow field as a solution to the fluid dynamic conservation laws. Consequently, the exit boundary condition was retained only at the downstream domain boundary, as shown in Fig. 3.4.

The effect of moving the outer boundary away from the wing may be seen in Table 3.1. The boundary was moved outward by adding additional grid points to the basic $5c \times 5c$ grid so that the outer boundary location is isolated from other grid-dependency issues. For all grids, 100 grid points were used along the wing chord. As may be seen from the table, there is very little

Table 3.1. *Effect of outer boundary location on computed aerodynamic properties of a rigid 2% circular arc airfoil at* $\alpha = 5°$, Re $= 4 \times 10^3$

grid	domain size	C_L	C_D	C_M
185×77	$5c \times 5c$	.637	.0761	.0442
207×91	$10c \times 10c$	.626	.0752	.0425
221×101	$15c \times 15c$	.623	.0750	.0422
247×121	$30c \times 30c$	.621	.0749	.0420

change–less than 1%–in the computed lift, drag, and aerodynamic moment beyond a domain size of $10c \times 10c$. Consequently, a computational domain size of $10c \times 10c$ with exit conditions imposed at the downstream boundary is adopted for all further computations.

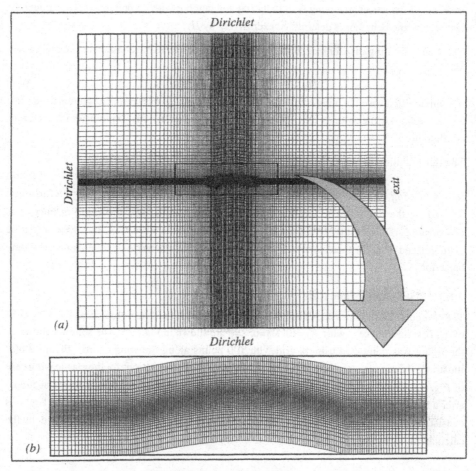

Figure 3.4   Typical grid used for membrane wing computations. The complete computational domain and boundary conditions are shown in (a), and an enlarged view of the grid near the membrane is shown in (b).

3.2.2 Classification of Flexible Membrane Wings

Membrane wings may be broadly classified by considering the following three limiting cases of the parameters Π_1, Π_2, and ε. The first limiting case may be referred to as the elastic wing case with

$$\Pi_1 \ \_ \ \text{finite}, \ \Pi_2 = 0, \ \varepsilon = 0 \qquad (3.26)$$

In this case the membrane wing is initially flat and taut but with no pretension. Consequently, the membrane behavior is determined exclusively by Π_1.

The second limiting case for membrane wings may be referred as the constant tension wing case with

$$\Pi_1 = 0, \ \Pi_2 \ \_ \ \text{finite}, \ \varepsilon = 0 \qquad (3.27)$$

As with the first case the wing is initially flat and taut but now has sufficient pretension so that the material stiffness does not participate in determining the equilibrium membrane configuration.

The third limiting case may be referred to as the inextensible wing case with

$$\Pi_1 \ \_ \ \text{infinite}, \ \Pi_2 = 0, \ \varepsilon \ \_ \ \text{finite} \qquad (3.28)$$

In this limiting case, the wing is not initially flat and taut but has an excess length of material defined by the parameter ε. Of course, the Reynolds number and the angle of attack are additional parameters that characterize a steady flow about an airfoil, but they are to be distinguished from the three parameters listed above that characterize flexible membrane wing behavior.

3.2.3 Elastic Membrane Case

The computed streamlines and constant pressure contours for an elastic membrane wing typical of the first limiting case with $\Pi_1 = 10$, $\Pi_2 = 0$, $\varepsilon = 0$ are shown in Fig. 3.5. The solution shown in the figure was computed using a grid with 100 points along the wing chord. Figure 3.6(a) illustrates the convergence path of the four coupled governing equations for the elastic wing case of Fig. 3.5. It can be seen that the residuals are reduced to single-precision machine accuracy after a few thousand iterations. Also, the terminal level of the residuals shown in the figure is consistent with the single-precision floating point accuracy of the arithmetic used in the calculation.

Figure 3.6(b) compares the computed membrane surface pressures for the case of Fig. 3.5 with the surface pressures predicted by a potential flow calculation for the same membrane configuration. The discrepancy between the potential solution and viscous flow solutions appears to be attributable to a general loss of circulation about the wing due to viscous effects

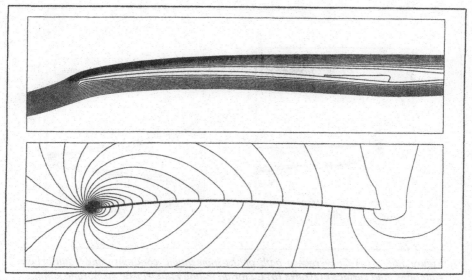

Figure 3.5   Streamlines and isobars for an initially flat elastic membrane wing.
Re = 4 x 10^3 , α = 6°, Π_1 = 10, Π_2 = 0, ε = 0.

primarily near the trailing edge. However, for this set of aeroelastic parameters, potential flow does reasonably well in predicting the membrane surface pressures. For other choices of the dimensionless parameters, potential flow is not always as successful in capturing the fluid dynamics of the problem.

3.2.4   Inextensible Membrane Case

The limiting case of a flexible but inextensible membrane wing, with excess length ε=.017, Π_1=46, and Π_2=0, is investigated in this section. The excess length and aeroelastic parameters were chosen to reproduce a portion of the experimental data reported by Newman and Low (1984). Experience has shown that for this excess length, any further increase in Π_1 will not substantially effect the aeroelastic solution. Consequently, the results presented are for a membrane wing that may be considered essentially inextensible. Due to the assumption of steady laminar flow, the numerical computations were limited to Reynolds numbers below 4x10^3, whereas the Reynolds number reported by Newman & Low (1984) was 1.2x10^5, which is in the turbulent flow regime.

The effect of grid refinement on the computed streamlines for the inextensible membrane wing at Re = 4 x 10^3 and 0° angle of attack is shown in Fig. 3.7. As may be seen in

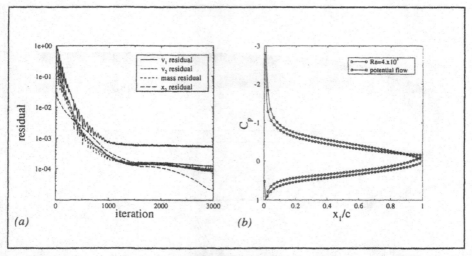

Figure 3.6.   (a) Convergence path of the momentum, continuity and equilibrium equations, and surface pressures (b) C_p for an initially flat, elastic membrane wing. Re = 4 x 10³, α = 6°, Π_1 = 10, Π_2 = 0, ε = 0. In the figure, v_1, v_2 are the momentum residuals, and x_2 is the residual for the equilibrium equation.

the figure, a large recirculation region appears on the lower surface of the membrane as the grid is refined. The integrated aerodynamic properties and the nominal membrane tension are given in Table 3.2 for the grid refinement sequence of Fig. 3.7. Even for the highest grid resolution shown, the tabulated values of lift and tension are not yet grid-independent. Of course, the membrane lift and tension will follow the same trend as the grid is refined, as required by equilibrium.

Figure 3.8 contrasts the viscous flow-based aeroelastic solution using 400 points along the wing chord with a potential flow-based aeroelastic solution for the inextensible membrane wing of Fig. 3.7. Although the equilibrium membrane profiles are similar, as shown in Fig. 3.8(a), the surface pressures, shown in Fig. 3.8(b), are dramatically different. In particular, near

Table 3.2. Effect of grid refinement on computed aeroelastic properties for an inextensible membrane wing at α = 0°, Π_1 = 46, Π_2 = 0, ε = 0.017, Re = 4 x 10³.

grid	points on c	C_L	C_D	C_M	C_T	max Y
261×101	100	.378	.0716	.152	.893	.0836
361×121	200	.342	.0689	.144	.823	.0835
461×141	300	.317	.0685	.139	.773	.0835
561×141	400	.300	.0702	.136	.741	.0834

the leading edge, the pressure jump across the membrane, Δp, predicted by the two flow models are of different sign. The negative Δp computed by the viscous flow model implies an inflection point in the equilibrium membrane profile which can be seen in the figure. In contrast, the potential-based model predicts a flow pattern and a membrane profile that is completely symmetric about the midchord. It can be concluded that for this set of dimensionless parameters, a potential flow-based membrane wing theory essentially fails.

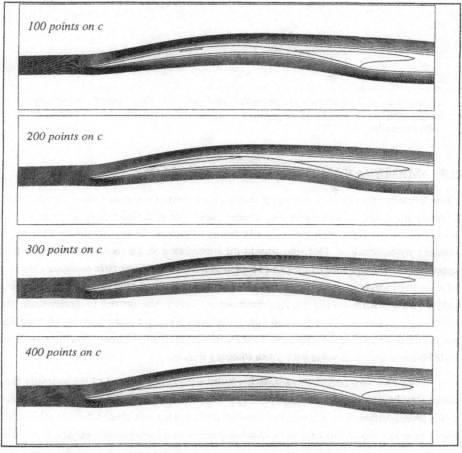

Figure 3.7   Effect of grid refinement on streamfunction contours for an inextensible membrane wing with excess length.

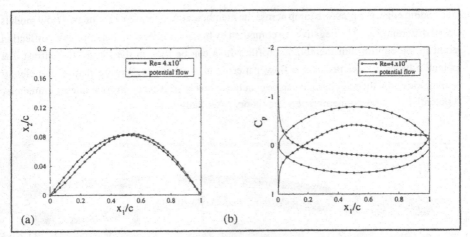

Figure 3.8   Comparison of potential and viscous flow based membrane configurations, (a), and surface pressures, (b), for an inextensible membrane wing with excess length.

3.3   MEMBRANE WINGS IN UNSTEADY FLOW

In the following, the role of viscosity in the unsteady flow over a membrane wing is investigated. In many situations, viscosity may play a pronounced role in the unsteady aerodynamics of membrane wings since the membrane will adjust to changes in the freestream velocity as required by equilibrium. This adjustment in the membrane shape may lead to the periodic appearance or disappearance of regions of separated flow. In addition to investigating the role of viscosity in the aeroelastic boundary value problem, the simulation of a membrane wing subjected to a harmonically varying freestream velocity also serves to demonstrate the differences between the three limiting aeroelastic cases discussed in the previous chapter. For each of the three limiting cases presented in the following sections, the freestream velocity varies by 20% about the mean value with a given forcing frequency.

The time step required to adequately resolve the major features of the unsteady flow about a membrane wing can be determined by comparing the aeroelastic solutions computed using several different time steps. The constant tension wing was chosen for this study since experience has shown this case to be quite responsive to a harmonically varying freestream velocity.

Figure 3.9 shows the time series of the aeroelastic solution using dimensionless time steps of 0.075, 0.15, and 0.30 with a 281×101 spatial grid. The freestream velocity, aerodynamic lift coefficient, and membrane profile coordinate at the midchord point for each of the three time steps are shown in the figure. The dimensionless parameters that govern the solution are given in the figure caption. It can be seen from Fig. 3.9 that even for the smallest time step the aeroelastic solution is not yet time step-independent. It is interesting to note that the solution is essentially independent of the time step during the portion of the cycle when the freestream is accelerating and, conversely, strongly dependent on the time step when the freestream is decelerating. The reason for this strong time step dependency during periods of freestream deceleration will become clear in the following section. On a more practical note, it is known from the previous section that the steady-state solution is not also completely grid-independent at this level of spatial resolution. Therefore, the spatial and temporal accuracy afforded by a 281×101 grid and a nondimensional time step of 0.075 is considered adequate for the present investigation and is adopted for all unsteady computations.

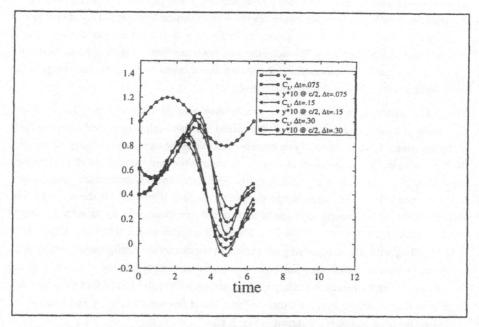

Figure 3.9   Variation in the computed aerodynamic lift and midchord coordinate using three different time steps for a constant tension membrane wing subjected to harmonic freestream oscillation. The nondimensional parameters are $\varepsilon=0$, Re=4 x 10^3, $S_t = 1.5$, $\Pi_1=0$, $\Pi_2=2$, and $\alpha=4°$.

3.3.1 Constant Tension Membrane Case

The first case to be investigated is that of an initially flat, constant tension wing. Figure 3.10(a) shows the time series of the freestream velocity, the aerodynamic lift, drag and moment, and profile coordinate at the midchord point for the membrane wing at 4 degrees angle of attack. The initial condition for this simulation is taken to be the steady-state solution to the aeroelastic problem with the same dimensionless parameters. Also, the time shown in Fig. 3.10(a) is the dimensionless time defined by Eq. (3.9c). Figure 3.10(b) and 3.10(c) show the membrane surface pressures and membrane profile, respectively, at several instants during the second complete cycle of the freestream velocity.

Several observations can be made concerning the solution shown in Fig. 3.10(a). First, it can be seen that the peak in the membrane deflection lags the peak in the freestream velocity by approximately 60°. This phase lag, as well as the large amplitude of the motion of the membrane–the membrane camber varies from negative 1% to positive 8%–is characteristic of a system that is being driven near, but below the system natural frequency. This observation is supported by the value of the frequency ratio Ω_2 (Eq. 3.18), which has a value that is approximately 1.0 for this case. Secondly, the membrane profiles in Fig. 3.10(c) show that the algorithm is capable of successfully simulating the dynamics of membrane wings when substantial changes in the membrane profile occur.

The variation in the membrane profile may also be seen in Fig. 3.11 where the streamfunction contours are shown at equally spaced intervals during the second complete cycle of the freestream velocity. In this figure it can be seen that as the freestream velocity decelerates, the flow separates along the upper surface of the membrane, and two regions of recirculating flow develop near the trailing edge. The periodic appearance and collapse of these recirculation zones, as seen in Fig. 3.11, suggests that the role of viscosity is enhanced in the unsteady flow scenario, since the acceleration and deceleration of the freestream velocity strongly influences the separation and reattachment of the flow. The periodic appearance and collapse of these flow features, along with an accompanying adjustment in the membrane configuration, results in an aeroelastic response which may not be characterized as simple harmonic response at the freestream forcing frequency. A leading edge separation bubble also can be seen to appear near the end of the freestream cycle and then collapse as the freestream velocity accelerates. The associated pressure contours are shown in Fig. 3.12.

3.3.2 Elastic Membrane Case

The second unsteady case to be investigated is that of an initially flat elastic wing. In this case the membrane develops tension as a result of material strain rather than pretension. The

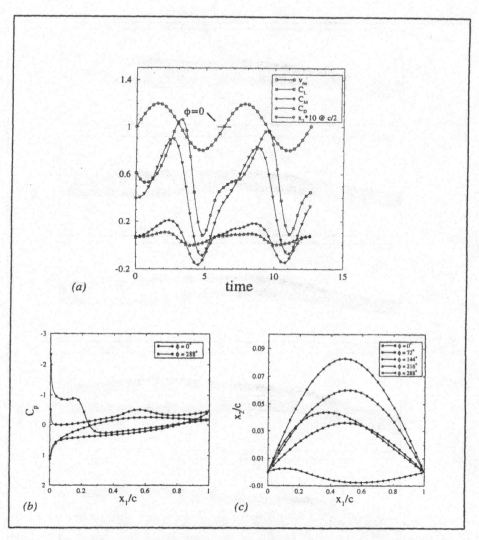

Figure 3.10 *Time series of aerodynamic lift, moment, and drag coefficient for a constant tension membrane wing subjected to harmonic freestream oscillation is shown in (a). Surface pressures and membrane profiles at several times during one complete oscillation of the freestream velocity are shown in (b) and (c), respectively. The nondimensional parameters are $\varepsilon=0$, Re=$4x10^3$, $S_t=1.5$, $\Pi_1=0$, $\Pi_2=2$, and $\alpha=4°$.*

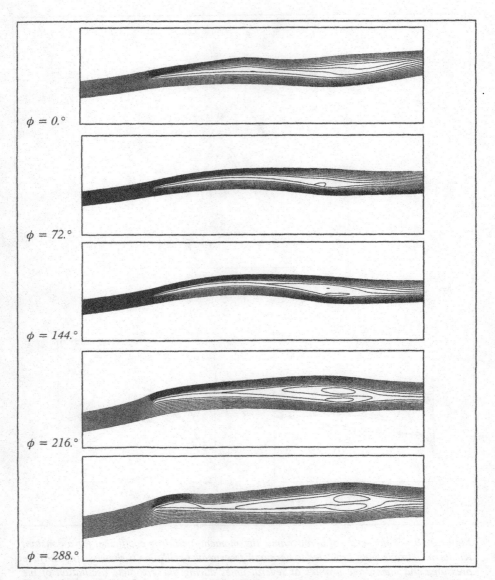

$\phi = 0.°$

$\phi = 72.°$

$\phi = 144.°$

$\phi = 216.°$

$\phi = 288.°$

Figure 3.11   Streamfunction contours at equally spaced time intervals during one complete oscillation of the freestream velocity for the case of the previous figure.

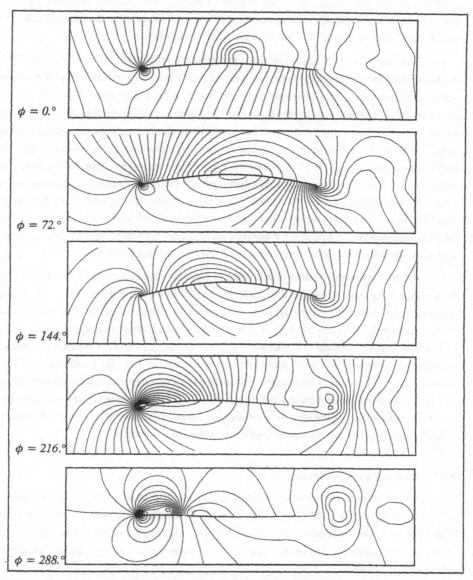

$\phi = 0.°$

$\phi = 72.°$

$\phi = 144.°$

$\phi = 216.°$

$\phi = 288.°$

Figure 3.12   Constant pressure contours at equally spaced time intervals during one complete oscillation of the freestream velocity for the case of the previous figure.

controlling aeroelastic parameter Π_1 was chosen so that the steady-state aeroelastic solution is the same as the constant tension case previously discussed. Here, as with the previous case, the steady-state solution was used to initialize the solution at time $t=0$.

The time series of the aerodynamic and membrane properties is shown in Fig. 3.13(a). Again, several observations can be made concerning this time series. First, the lift and membrane deflection show very little deviation from the steady-state value when compared to the deviations observed in the constant tension case. The difference between the two cases may be explained by recalling that the deflection of an initially flat, elastic membrane without pretension is proportional to the cube root of the applied pressure. Consequently, an elastic wing will become progressively stiffer as the freestream velocity is increased. This effect usually is referred to as geometric or stress stiffening. In this case, the effect of geometric stiffening serves to restrict the development of wing camber, and consequently, the appearance of regions of separated flow is less pronounced than in the constant tension case of the previous section. The membrane deflection, as well as the membrane aerodynamic properties, also can be seen to be an essentially simple harmonic response at the freestream forcing frequency.

Another interesting feature of the elastic wing case is the apparent lack of a phase shift between the freestream velocity and the membrane deflection at the midchord. This situation is consistent with a system that is being driven well below its natural frequency. This observation is supported by the frequency ratio Ω_1 (Eq. 3.18), which for this problem is approximately 0.1. The streamfunction contours for this case are shown in Fig. 3.14 and the associated pressure contours are shown in Fig. 3.15. It can be seen by comparing Fig. 3.10 and Fig. 3.13 that although the two cases have nearly identical steady-state characteristics (recall the solution at $t=0$ is the steady-state solution in both figures), the unsteady responses of the two problem classes is dramatically different. The differences between the two cases may be broadly attributed to the geometric nonlinearity inherent in the elastic problem.

3.3.3 Inextensible Membrane Case

The final class of problems to be investigated is the case of an elastic wing with the material stiffness tending to infinity. The controlling dimensionless parameter ε, was chosen to coincide with a portion of the experimental data reported by Newman and Low (1984). The value of the aeroelastic parameter Π_1 was chosen large enough so that the membrane may be considered essentially inextensible.

Figure 3.16(a) shows the time series of the aerodynamic and membrane properties for the inextensible wing at zero degrees angle of attack. It can be seen from the figure that the profile

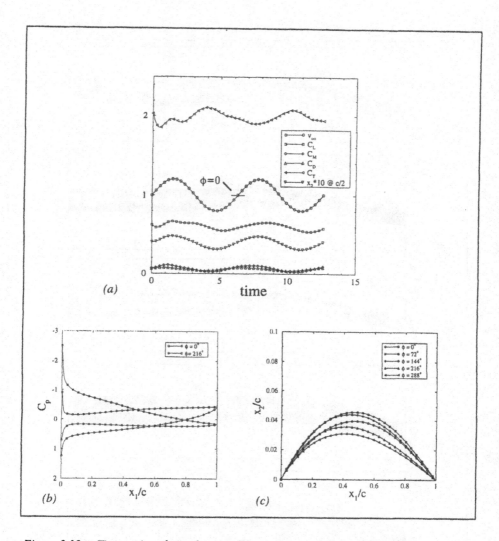

Figure 3.13   Time series of aerodynamic lift, moment, and drag coefficient for an elastic membrane wing subjected to harmonic freestream oscillation is shown in (a). Surface pressures and membrane profiles at several times during one complete oscillation of the freestream velocity are shown in (b) and (c), respectively. The nondimensional parameters are $\varepsilon=0$, $Re=4x10^3$, $S_r=1.5$, $\Pi_1=7.9$, $\Pi_2=0$, and $\alpha=4°$.

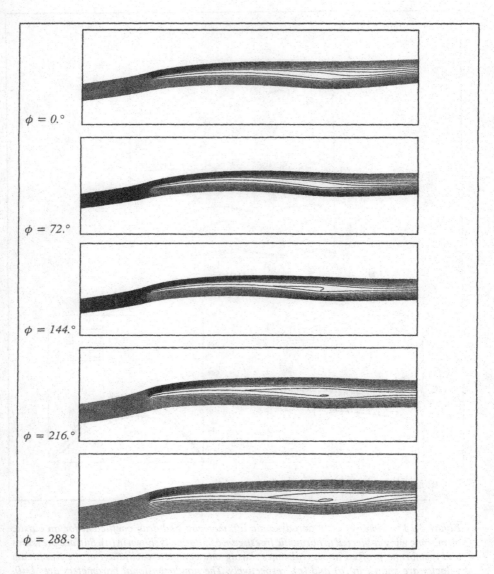

Figure 3.14   Streamfunction contours at equally spaced time intervals during one complete oscillation of the freestream velocity for the case of the previous figure.

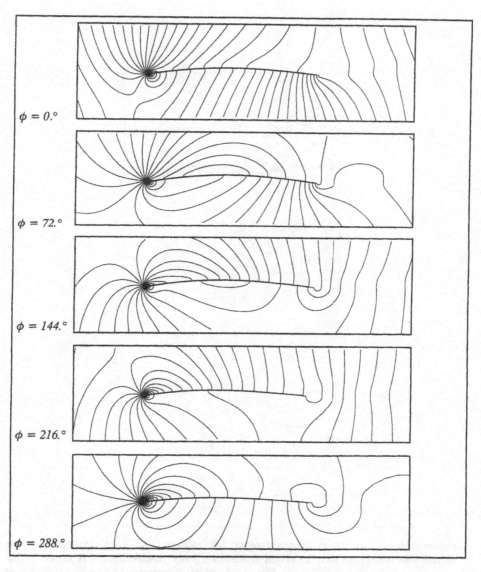

$\phi = 0.°$

$\phi = 72.°$

$\phi = 144.°$

$\phi = 216.°$

$\phi = 288.°$

Figure 3.15   Constant pressure contours at equally spaced time intervals during one complete oscillation of the freestream velocity for the case of the previous figure.

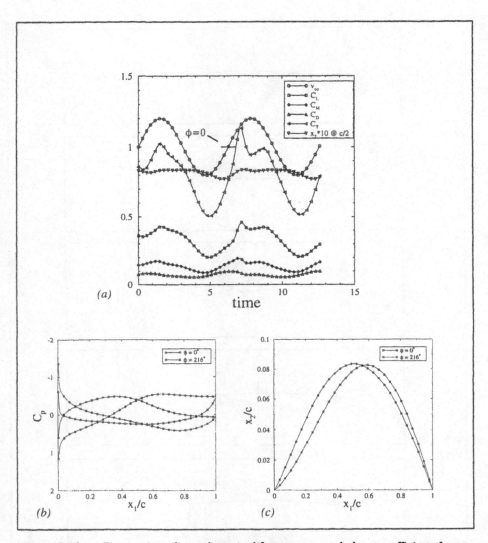

Figure 3.16.   Time series of aerodynamic lift, moment, and drag coefficient for an inextensible membrane wing subjected to harmonic freestream oscillation is shown in (a). Surface pressures and membrane profiles at several times during one complete oscillation of the freestream velocity are shown in (b) and (c), respectively. The nondimensional parameters are ε=0.017, Re=4x10³, S_r=1.5, Π₁=46, Π₂=0, and α=0°

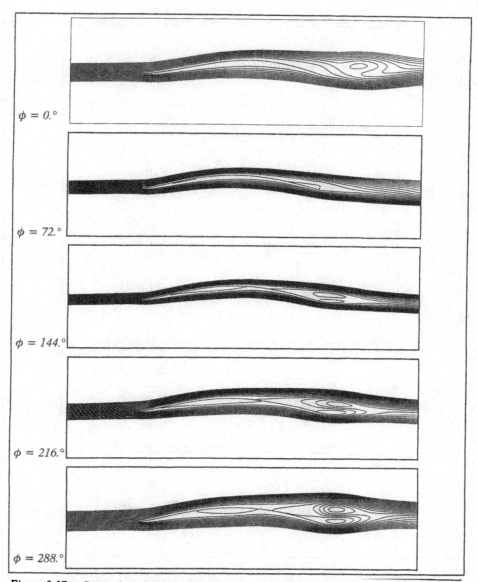

Figure 3.17. *Streamfunction contours at equally spaced time intervals during one complete oscillation of the freestream velocity for the case of the previous figure.*

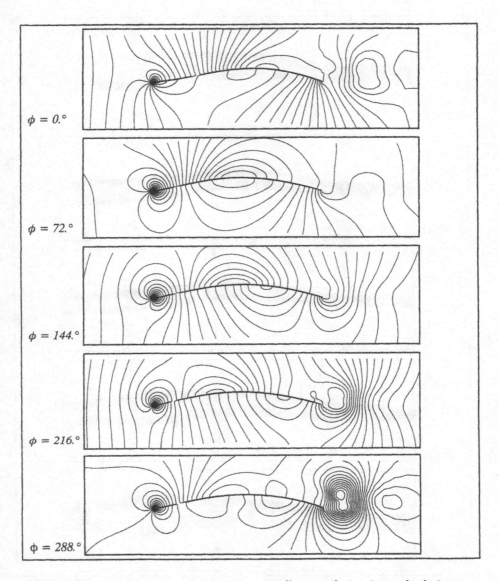

$\phi = 0.°$

$\phi = 72.°$

$\phi = 144.°$

$\phi = 216.°$

$\phi = 288.°$

Figure 3.18.   Constant pressure contours at equally spaced time intervals during one complete oscillation of the freestream velocity for the case of the previous figure.

coordinate of the membrane at the midchord point is largely unaffected by variation in the freestream velocity. However, the membrane profile does change shape in response to the unsteady flow field. This shape change is shown in Fig. 3.16(c) at two time instants during the second complete cycle of the freestream velocity. For this class of problems, the membrane profile is responding primarily to the periodic appearance and collapse of regions of recirculating flow on both the upper and lower surfaces of the membrane. These regions of recirculating flow can be seen in Fig. 3.17. As the freestream begins to decelerate, a region of separated flow appears on the lower surface of the membrane and a pair of counter-rotating recirculating zones appears on the upper surface of the membrane near the trailing edge. As the freestream velocity begins to accelerate, these recirculating zones quickly collapse, and the flow once again becomes fully attached. As was the case with the constant tension wing, the aeroelastic response shown in Fig. 3.16 may not be characterized as simple harmonic response at the freestream forcing frequency. The associated pressure contours are shown in Fig. 3.18.

3.4   SUMMARY AND CONCLUSION

In this chapter, a numerical model simulating the aeroelastic characteristics of a flexible two-dimensional membrane wing has been presented. The use of the Navier-Stokes equations in the present model is a substantial departure from previous work on membrane wing aerodynamics which has, almost universally, adopted a potential-based description of the flow field. The two-dimensional aeroelastic boundary value problem was introduced and nondimensionalized and a set of six basic dimensionless parameters was derived which govern the solution of the problem. An additional parameter, the frequency ratio, was proposed as a meaningful parameter for characterizing the harmonically driven unsteady aeroelastic problem.

A moving grid procedure was then applied to the coupled aeroelastic problem and was shown to yield results that are in agreement with available analytic solutions for several appropriate limiting cases. The numerical procedure also was shown to satisfy certain identities as dictated by the fundamental fluid dynamic conservation laws. The role of viscosity in membrane wing aerodynamics was investigated using the numerical model for both steady and unsteady flows. These investigations were facilitated by distinguishing three classes of problems which are associated with three limiting cases of the dimensionless parameter set. The aerodynamic characteristics of membrane wings at Reynolds numbers between 10^3 and 10^4 were shown to be substantially different from those predicted by a potential-based membrane wing theory. The role of viscosity was shown to be preeminent in the harmonically forced unsteady flow about a membrane wing. In this case, the influence of viscosity is enhanced since the acceleration and deceleration of the freestream velocity strongly influences the separation and reattachment of the flow. The periodic appearance and collapse of recirculation zones, along with an attendant adjustment in the membrane configuration, results in an aeroelastic response which may not be characterized as simple harmonic response at the freestream forcing frequency. The whole subject of fluid–structure interaction has been under intensive research recently. A summary of an interesting body of work using adaptive unstructured grids along with some impressive results can be found in Löhner et al. (1995).

MOVING GRID TECHNIQUES:
MODELING SOLIDIFICATION PROCESSES

4.1 INTRODUCTION

In the previous chapter, we applied a boundary-fitted grid approach to an aeroelastic problem. This same methodology also can be applied to the evolution of a phase-change interface. In the aeroelastic problem, the membrane has fixed ends with respect to the incoming flow, whereas in the phase change problem, the interface translates across the domain, replacing one phase with another. Furthermore, in the aeroelastic problem, the deformation of the moving boundary was limited by the end conditions and the elasticity of the membrane. In the phase-change problem, the shape and the motion of the interface is determined only by the heat flux jump across the interface. Thus, the demands on the numerical algorithm for front tracking may be much heavier than for the membrane problem. In order to gain an appreciation of these demands, we briefly introduce the physical conditions leading to the instability phenomenon in solidification.

4.1.1 Morphological Instabilities During Solidification

To illustrate some of the key issues associated with moving boundary problems in the context of phase-change, we choose as our physical system the growth of crystals of pure material from a melt. Under certain growth conditions–conditions not infrequently encountered in conventional solidification processes–the solid-liquid interface is likely to become unstable, resulting in highly branched morphologies. The mechanisms of instability and various aspects of pattern selection and transport processes have been thoroughly investigated over the years, and the literature abounds in references (Brenner 1991, Huang and Glicksman 1981, Langer 1981, Langer and Muller-Krumbhaar 1978, Mullins and Sekerka 1964, Pelce 1988, Shyy 1994). Here, we detail only such features of the morphological evolution as are necessary to provide a motivation for the numerical procedure presented later.

Morphological instabilities at a solid-liquid or solid-vapor phase boundary determine the shape and properties of grown crystals. An inspiring and perhaps the most familiar example of the consequences of such instabilities is the snowflake, which assumes a highly branched, six-fold symmetric shape under certain growth conditions (Nakaya 1954). As early as in the seventeenth century Kepler (Emerton 1984, p. 41) observed that

> *The cause of the six-sided shape of a snowflake is none other than that of the ordered shapes of plants and of numerical constants. . . . I do not believe that even in a snowflake this ordered pattern exists at random. . . .There is, then, a formative faculty in the body of the earth. . . . The faculty of earth is in itself one and the same, but it imparts itself to different bodies and cooperates with them. It engrafts itself on them and builds now one design, now another, as the inner disposition of each matter or outer conditions allow.*

Although Kepler's insights into the mechanisms of formation of the snowflake were constrained by an inadequate knowledge, of the states and fundamental structure of matter, he correctly anticipated some of the key features of pattern-forming processes. In particular, his intuition regarding the interplay of microscopic (inner disposition) and macroscopic (outer conditions) features has been borne out by research in the past few decades (Kessler et al. 1988). The scaling aspects of the inner and outer conditions will be addressed later in this and subsequent sections.

For some time now, quite successful efforts have been directed at unraveling the physics underlying such repeatable patterns as the snowflake. The motivation is to unearth a paradigm for organized phenomena far from equilibrium. On the most fundamental level, one is interested in revealing the causes and effects of the plethora of patterns observed in nature. Biological systems, for example, abound in forms driven by function. In a recent review of the current state of knowledge in the area of pattern formation, Cross and Hohenberg (1993, p. 1050) write, in connection with biological patterns,

> *. . . at the earliest stages of development there arises the question of how spatial differentiation arises in a featureless medium. A mathematical form of this question concerns the breaking of a symmetry: one end of a symmetric egg will become the head and the other the tail. As with any broken symmetry, chance, i.e., fluctuations, can determine which is which, but this choice is only binary–the head will not appear in the middle of the embryo. As the organism grows, different parts will develop quite different*

functions, despite the fact that the material of the embryo is rather homogeneous, at least at the level of the molecular chemistry controlling development. It is true that on its full scale, the embryo is not spatially uniform due to its finite size, and ultimately this limitation will play a vital role. The question remains of how this macroscopic nonuniformity is communicated to the smaller scales.

It is obvious that the issues pertaining to the formation of structures in biological systems are considerably more complex than in other physical systems. The fundamental physical processes and principles underlying biological patterns are not well understood. The causes may be biochemical in origin and may involve environmental effects. In contrast, the solidification process represents one of the simpler systems displaying pattern formation and has been widely researched for that reason. Nevertheless, even in this system several unanswered questions remain. The numerical technique detailed here is directed towards answering some of these questions.

The interest in pattern-forming phenomena during solidification does not stem solely from aesthetic or theoretical standpoints. Strong technological motivations exist for the study of such patterns (Brown 1988, Flemings 1974). For example, the demands on purity and crystal quality of semiconductors and high-temperature alloys are becoming ever more stringent. Under commonly encountered solidification environments employed in the preparation of such materials, instabilities can occur causing the solid-liquid interface to break up into the morphologies shown in Fig. 4.2. When such morphologies result, the solutes or dopants are trapped in the interdendritic or intercellular regions (microsegregation) leading to compositional inhomogeneities. The nonuniformities can be discerned from the characteristic patterns observed in metallographic sections. They determine the grain sizes, defects, and other crystal properties. To mitigate the impact of instabilities, materials often are processed at rates, solute concentrations, and temperature levels at which they can be avoided or reduced. In producing alloys, for example, instabilities at the interface can be avoided at low growth rates, while maintaining large temperature gradients in the melt. Thus, the process variables and economies of solidification processes are influenced by considerations involved in dealing with morphological instability. While maintaining low growth rates may conflict with the economics of production, large temperatures may lead to thermal stresses that may influence the structural properties of the crystals produced. Therefore, much attention has recently been directed by the materials science community to fully understand the instability phenomena and develop the capability to predict and control them. The complexity of the physics involved in these highly nonequilibrium, nonlinear phenomena is further compounded by the microscopic scale of events, the high temperatures at which metallic systems operate, and by the opacity of metallic melts. These problems hinder experimental as well as analytical investigations. Furthermore, as

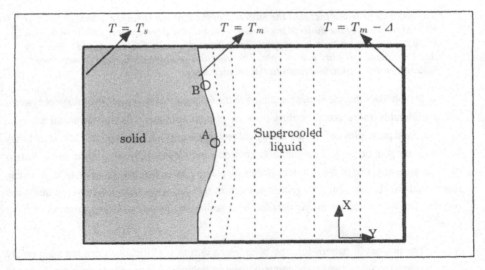

*Figure 4.1. Illustration of the Mullins–Sekerka instability mechanism.
Dotted lines represent isotherms*

will become clear subsequently, a plethora of physical issues enter into the overall solidification process, each with characteristic length and time scales and physical mechanisms predominant at these scales. Thus, in order to fully understand the solidification process and the pattern-forming phenomena involved therein, numerical simulation is perhaps the most viable investigative technique.

4.1.2 Physics of Morphological Instabilities in Solidification

Consider the typical crystal growth situation depicted in Fig. 4.1. Let the wall on the liquid side be maintained at a temperature T_l such that $T_l < T_m$, the melting temperature of the material, that is the melt is *undercooled*. The interface between solid and liquid corresponds approximately to the isotherm $T(x,t)=T_m$, after including the local curvature effect expressed in the Gibbs-Thomson condition (Langer 1980). Thus, the protrusion of the bump into the liquid leads to a clustering of the isotherms in the vicinity of the tip at the point A, implying a higher temperature gradient there. The bump thereby is induced to lose latent heat more rapidly in comparison with the other regions of the interface, such as B. This causes the disturbance to run away to form a finger. For the growth of an unstable front in a pure material system, the available exact solutions were obtained by Ivantsov (Huang and Glicksman 1981, Langer 1980, Pelce 1988) in the form of paraboloids of revolution. These solutions have the property that for given values of undercooling, they represent a continuous family of solutions and specify only the combination $\varrho_t V_t$, where ϱ_t is the tip radius and V_t is the tip velocity of the paraboloid. However, it is observed that the tip radius and velocity assume unique values as functions of undercooling in real growth systems (Huang and Glicksman, 1981). Thus, a selection mechanism is missing from the Ivantsov model, which failed to take account of an important physical ingredient, namely surface tension. Surface tension appears as the controlling factor to select a discrete set of solutions from the continuum of solutions (Kessler et al. 1988). Surface tension also stabilizes the interface by modifying the interfacial temperature, which no longer corresponds to the T_m isotherm associated with a planar interface. In particular, the Gibbs-Thomson condition for the interfacial temperature T_i, reads:

$$T_i = T_m \left(1 - \left(\gamma(\phi) + \frac{\partial^2 \gamma(\phi)}{\partial \phi^2} \right) \frac{\varkappa}{L} \right) \tag{4.1}$$

where $\gamma(\phi)$ is the surface tension, L is the latent heat, ϕ is the angle between the normal and the x-axis, and $\varkappa$ is the interfacial curvature (Woodruff 1973). Thus, surface tension depresses the temperature at the tip, reducing the effective undercooling there and providing a stabilizing effect. In essence, surface tension implies the existence of a critical nucleation radius and provides a short wavelength cutoff for the instability. The selection of a unique dendrite tip, from the bounded continuum caused by surface tension, is accomplished by considering anisotropy of the crystal lattice, as represented by the angle ϕ in Eq. (4.1). The selected dendrite tip, however, is highly sensitive to noise and behaves like an amplifier (Pieters and Langer 1986). Thus, the final shape of the crystal usually is not represented by a smooth paraboloid, but a highly branched structure, with successive instabilities ensuing on side-branches at different scales (Huang and Glicksman 1981). A schematic of this shape is displayed in Fig. 4.2(a), along with a cellular structure usually arising in the solidification of alloys (Fig. 4.2(b)). Beyond the branched structure, other events result. The final complete solid is obtained by a *coarsening* and *ripening* process. When the entire system cools, sidebranches merge, and a highly interconnected solid structure forms with isolated liquid pools that also eventually solidify (Brown 1988, Flemings 1974, Rappaz 1989). Besides the thermodynamic requirement of minimization of Gibbs' free energy, the physics underlying such events is yet to be characterized theoretically.

   Although the idealized diffusion-controlled free dendritic growth process has offered considerable insight into the physical mechanisms of solidification, it is accepted that in any realistic solidification processing arrangement it is impossible to avoid some form of convection (Brown 1988, Favier 1990, Glicksman et al. 1986, Ostrach 1983). Fluid flow can arise in a

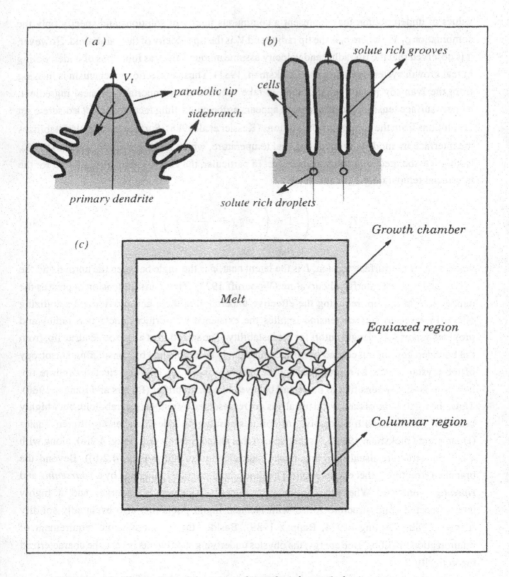

Figure 4.2. Illustration of the types of interfacial morphologies upon instability at the solid–liquid interface. (a) Dendritic and (b) cellular morphologies arising during solidification. (c) Columnar and equiaxed crystals in solidification from the melt. The envelopes of the crystals are shown in both cases.

variety of ways. Consider some typical solidification processes shown in Fig. 4.3. In these crystal growth techniques, capillarity effects at the free surfaces lead to Marangoni convection, and the pulling of the crystal leads to forced flow. Usually the crystal also is rotated in order to enforce uniformity along the circumferential direction, again leading to forced convection. Additionally, in the presence of gravity, natural convection can occur due to the large temperature gradients imposed in the axial as well as radial directions. The macroscopic solid-liquid interface therefore evolves under the influence of various forms of convection, whose strengths and flow features can vary depending on growth conditions. In growing crystals of a desired homogeneity, one is concerned with precise control of the transport phenomena. In particular, the distribution of impurities can be significantly affected by the presence of interfacial instabilities due to solute being trapped in regions between the cellular and dendritic structures illustrated in Fig. 4.2. Such microsegregation can be influenced by convective effects. In addition, the impurity and temperature distribution in the vicinity of the interface is a function of the global flowfield and heat transfer characteristics. It therefore is clear that crystal growers have to contend with highly inhomogeneous unsteady growth environments and flowfields. In maintaining the uniformity of grown crystals, it is important to minimize and/or control the effects of convection at the global and morphological levels. For these reasons, considerable interest has developed regarding the effects of convection–both natural and forced–on the macroscopic and microscopic interfacial characteristics (Shyy 1994). This interest has been motivated by low-gravity crystal growth experiments (Ostrach 1983), and the possibility of controlling microstructure using forced convection (Bouissou et al. 1990). In a system already complicated by the phenomena described, the addition of convective effects further obfuscates the issues by bringing in length and time scales disparate from the instability phenomena, and breaking symmetry. Theoretical investigations thus far have been only of an idealized nature (Ananth and Gill 1989, 1991, Davis 1990).

Based on research aimed at the fluid dynamic aspects of solidification, there is now abundant evidence relating the effects of convection in the melt to the characteristics of the phase interface. For example, Fang et al. (1985) have experimentally observed a strong coupling of crystal-melt interface shape with convection in a long circular cylinder of pure Succinonitrile, an organic FCC crystal. The experiments of Gau and Viskanta (1986) on the melting of Gallium clearly demonstrate the effects of natural convection on the deformation of the macroscopic solid-liquid interface. Depending on the strength of convection, measured by the Rayleigh number, the interface can be severely distorted by the flow pattern. Several computational efforts have attempted to reproduce the experimental results, with varying degrees of success (Lacroix 1989, Lacroix and Voller 1990). As exemplified by a recent experiment (Campbell et al. 1994), there are difficulties associated with identifying the interface in an opaque metallic melt with

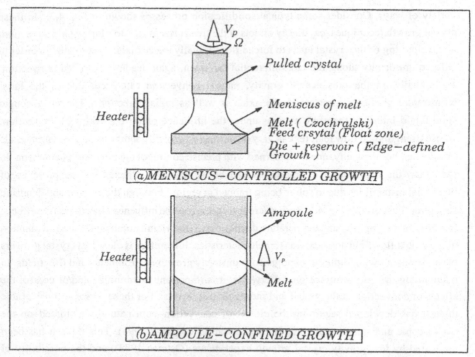

Figure 4.3. Two examples of solidification processing arrangements. Different types of convection are possible, e.g., natural convection, forced flows due to pulling and rotation of crystal and shrinkage, and Marangoni convection. (a) Meniscus-controlled growth. (b) Ampoule-confined growth.

high melting points, and with ambiguous initial conditions. Accordingly, low melting point, transparent materials often are favored by experimentalists to study phase change dynamics. However, while for diffusion-controlled environments it is possible to use these transparent, low latent heat of fusion organic melts as replacements, it is not clear that such is the case when convection is included. This is because the Prandtl numbers ($Pr = \mu C_p / k$), which define the relative importance of momentum to heat diffusion rates, are vastly different for metals as compared to organic melts. Thus, the flow field in each case is likely to be very different, and it is no longer valid to extrapolate results regarding flow features and their effects from organic to metallic melts. Thus, even in the case of the macroscopic flow field, reliable experimental results for validation of solution techniques are scarce.

4.1.3 Implications of Morphological Instabilities

Thus far we have summarized the current state of knowledge of dendritic growth in pure materials where the diffusion of heat controls the instability phenomena. In alloys and semiconductors, which are grown by a variety of processes, the diffusion of solute is the limiting process, and cellular and dendritic forms are observed on the growth front. The physics involved in these cases is analogous to that described above. In the event of other control parameters entering the process by which solidification is obtained, such as in directional solidification, the instability phenomenon may become more complex (de Cheveigne et al. 1990). The origin of cellular forms in directional solidification is subcritical in origin, but it derives from the same mechanism as in the pure material case–the propagation of the solidification front into a metastable supercooled melt. For commonly employed alloys, the phase diagram is such that the solid rejects solute into the liquid phase. In particular, there is an accumulation of such excess solute in the intercellular grooves and ahead of the growth front, leading upon complete solidification to an inhomogeneous solute distribution. In addition, instabilities at the root of the cells may lead to the periodic shedding of droplets (Brattkus 1989, Kurowski et al. 1989), as shown in Fig. 4.2(b), resulting again in non-uniform solute distribution. At higher growth rates the cellular forms may transition into dendritic structures. The mechanism for this transition is not well understood.

Even in the case of pure materials, the effects of convection on dendritic growth have not been well established, and most models rely on purely diffusive field equations, although it has been demonstrated that both natural and forced convection in the melt can influence the microstructure (Ananth and Gill 1989, 1991, Bouissou et al. 1990, Fang et al. 1985, Ostrach 1983, Tirmizi and Gill 1987). The entire theory of pattern formation in solidification assumes that the interface is in local equilibrium, i.e., it ignores kinetic effects. The parameters associated with the kinetics of interfacial growth, such as rate constants, interfacial energies, anisotropies, and so on, are difficult to measure or model; these mechanisms have been largely ignored under the assumption of slow-moving interfaces. Other effects that may have significant impact on pattern selection in solidification include effects of confinement (Cladis et al. 1990, Trivedi and Somboonsuk 1984), geometry of the growth domain (Fabietti et al. 1990), and competition between orientations of heat transfer and crystalline anisotropy (Borisov et al. 1991).

Fig. 4.2(c) shows that dendrites can assume columnar or equiaxed forms, depending on their origin, that is, depending on whether nucleation takes place on a substrate such as a wall, or in the bulk of the melt. A metallographic section can display distinct regions of grains resulting from columnar and equiaxed growth and a transition between the two types. It has been hypothesized that dendrite branch detachment from the columnar region and subsequent convection and growth in the melt can promote equiaxed growth (Beckermann et al. 1994). Beyond the formation of mature cells and dendrites several changes take place, culminating in complete solidification. In particular, in the late stages of dendritic growth, a phenomenon of *coarsening* or *ripening* occurs whereby the length scales of the solid structure increase (Huang and Glicksman 1981). For example there is a progressive increase in sidebranch spacing. Sidebranches in the vicinity of the tip of the primary dendrite are closely spaced, while the sidebranch spacing is greater away from the tip. This coarsening of the sidebranch structure can occur via branch competition, whereby a sidebranch that is fortuitously longer than its neighbor suppresses the growth of the latter. Alternatively, two neighboring sidebranches can merge with the same result. All these phenomena are of utmost importance in determining the microstructures. In addition, it is important to determine the length and time scales involved. In particular, one needs to be able to predict the final primary dendrite spacing, length and velocity scales, the sidebranch spacing, and its correlation if any to the primary dendrite characteristics. These determine the dimensions of the envelope of the entire intricate branched structure corresponding to each dendrite. This envelope forms a coherent, correlated structure and defines the all-important grain shapes and sizes of the solidified material. Microsegregation– accumulation of impurities/solutes in the interdendritic regions–leads to a pattern of solute inhomogeneity upon formation of the final grain structure. The grain structure and resulting solute distribution govern the properties and performance of the metallic alloys or semiconductors.

The above considerations significantly influence the processes employed to produce crystals and have prompted single crystal growth techniques (Brown 1988) and efforts aimed at mitigating the instabilities at the solid-melt interface. A desire to avoid instabilities may influence processing parameters that may conflict with the economy of the solidification processes. There has thus been extensive effort to determine the effects of processing environments on the microstructure and vice versa (Beckermann and Viskanta 1993, Ni and Beckermann 1993, Rappaz 1989, Thevoz et al. 1989, Tseng et al. 1991). With the computational facilities now available, the governing equations at the two levels are easily solved. The difficulty lies in modelling and computing the interactions of the different scales. It is a distinguishing feature of the solidification morphology problem that a plethora of scales are present from the atomic scales ($O(\text{Å})$) to that of the growth environment ($O(\text{cm})$). In the solution of the macroscopic equations, it is necessary to model the properties and behavior of the two-phase zone, often called the *mushy zone* (Voller and Prakash 1987). This region is comprised mainly of the morphological structures and adjoining liquid phase, and there appears to be no alternative but to perform averaging operations to model it. Even so, it is necessary that the scales, segregation patterns and solid fractions associated with each dendrite/cellular structure are quantified. The coupling between micro- and macroscopic phenomena affect all these quantities. Investigation of the effects of macroscopic phenomena on the microstructure, such as convection, heat flow, and boundary conditions, on the microstructure, whether experimental or analytical, have been of a preliminary nature. Among efforts aimed at unifying the phenomena at different scales, the volume-averaged models of Beckermann and coworkers (Ni and Beckermann 1993, Wang and Beckermann 1993) and the solute diffusion models (Rappaz and Thevoz 1987, Thevoz et al. 1989) are notable. These models incorporate solidification phenomena from the stochastic nucleation process to the macroscopic features of the solidified product. However, the details at each scale are either modeled or adopted from empirical correlations. The difficulties involved with the multi-scale nature of the phenomena and the plethora of nondimensional numbers which influence the physical phenomena at the different scales have thwarted attempts at arriving at an accurate representation of the physics. What is needed is the separate treatment of each scale and then an investigation of their interactions. Such a treatment of the overall solidification phenomena is possible only with a systematic approach of constructing computational tools capable of solving the governing equations at all scales, by incorporating physics appropriate at each scale. The methods presented in this work, when appropriately combined, can lead to such a comprehensive computational framework.

4.1.4 Need for Numerical Techniques

To date, several instability phenomena involving interfaces have been studied by researchers from a wide range of specializations. While the snowflake is an example of a solid-vapor system that has experienced a *morphological instability*, the growth of ice crystals (Tirmizi and Gill 1987), in a solid-liquid system also gives rise to complex morphologies. Other well-investigated examples of interfacial instabilities include the Saffman-Taylor fingering phenomenon (Bensimon 1986, Saffman and Taylor 1958) between two fluids of different viscosities. Such fingers have been studied in connection with *water coning* in oil recovery (Glimm et al. 1988), flow through porous media (Homsy 1987), and in two-dimensional form in Hele-Shaw cells in the laboratory (Tabeling et al., 1987). The Saffman-Taylor instability is, in fact, an analogue of the solidification morphology problem, and several of the conclusions drawn from analyses of the former phenomenon have been found to apply to the latter (Kessler et al. 1988). Fluid-fluid systems such as liquid crystals and organic materials and the process of electrodeposition also are subject to interfacial instability in a similar fashion to the solid-liquid systems (Flesselles et al. 1991). Other interesting examples of interfacial instabilities in hydrodynamics include the Kelvin-Helmholtz instability in its various forms (Huerre 1987), the Rayleigh-Taylor overturning instability (Sharp 1984) and the breakup of liquid jets (Drazin and Reid 1981).

Conventionally, the onset of and linear growth stages of disturbances in the above mentioned phenomena have been investigated along the same lines as Lord Rayleigh's approach to the jet instability problem (Rayleigh 1899), that is, by applying linear stability theory. Weakly nonlinear analyses also have been performed (Langer 1980). However, far from onset, the interfaces undergo successive instabilities of various types, and in some systems the final observed pattern may deviate significantly from that predicted by linear stability analyses. In other cases, the instability may be triggered by large amplitude disturbances, and such subcritical phenomena obviously are not accessible to linear stability considerations. The mechanisms by which the nonlinear phenomena induce reorganizations and coherent patterns in unstable systems therefore are not clarified by analyses restricted to small perturbations. Thus, sufficient motivation exists for developing general investigative capabilities in order to study the above mentioned phenomena in their highly nonlinear stages. This regime is obviously the most accessible to laboratory and computational experiments.

The effort presented herein is devoted to developing a tool for computational study of the instability of interfaces far into the nonlinear growth stages. We choose to investigate the phenomenon of morphological instabilities during solidification due to its significance both as a typical pattern-forming system and in materials processing technologies. In addition, the instabilities in this system result in interfacial patterns, such as crystal morphologies, which pose a challenge to any simulation procedure. The interaction of the solidification front with itself, or with other fronts, and with the growth environments are typical of pattern-forming phenomena. Numerical simulation of the various events leading to complete solidification is evidently a complex task.

The growing demand for high-quality alloys and semiconductors calls for a means to predict and possibly control the distribution of impurities and additives in grown crystals. Instabilities at the solid-liquid interface can lead to patterns in the deposition of impurities (microsegregation) that may be detrimental to the performance of the materials solidified. It is important to be able to estimate such inhomogeneities in crystals and their relation to the macroscopic flow features and solute concentration. The subject of convective influence on cellular/dendritic growth is an open area of research. Numerical simulation has not been widely reported to date. In what follows, we attempt to commence the simulation of the full crystal interface evolution problem with melt flow, for shapes that are not constrained to be of small distortion. We develop a numerical simulation technique that possesses features needed to handle a typical solidification process, viz.

1.   The solid-liquid interface is a phase discontinuity; boundary conditions in each

phase have to be applied for the fluid flow equations on this moving internal boundary. The interface is a source of heat, mass, and momentum and can be arbitrarily distorted.

2. The method developed should be able to handle fluid flow in solidification processes at the macro and microscales.

Furthermore, computational issues such as conservation, accuracy, efficiency, and robustness of the chosen algorithm are to receive due consideration.

4.2 REQUIREMENTS OF THE NUMERICAL METHOD

Having laid out the physics of some typical morphological instability phenomena, we can now characterize the demands on the numerical techniques employed as follows:

1. The interface grows in perimeter in the course of the instability. Additionally, repeated branching results in a highly convoluted shape. Topological changes may result (Sato et al. 1987). Interfaces may merge or fragment. The solution procedure therefore is required to follow the evolution of an interface under these circumstances.

2. The behavior of the interface is very sensitive to the intricate details of the boundary conditions applied at the interface. In particular, as we demonstrate later, surface tension is a very delicate mechanism which subtly influences the final interface shape. Since surface tension multiplies the curvature of the interface, the numerical scheme is required to compute the interface shape and the first and second derivatives of the interfacial curve accurately in order to faithfully represent the physics.

The interface velocity is given, in the case of solidification, by the Stefan condition:

$$\varrho L v_n = (\ k_s \nabla T_s - k_l \nabla T_l \ ) \cdot \vec{N} \tag{4.2}$$

Here, ϱ, L, v_n, and k respectively are density, latent heat, normal velocity of the interface and thermal conductivity, $\vec{N}$ is the normal to the interface, T is the temperature, and the subscripts l, s represent liquid and solid phases. This is, in fact, the statement of conservation of heat for a control volume positioned at the interface. Thus, unless a field equation solver is developed to enforce this condition strictly at the interface, the interfacial velocity will be inaccurately obtained. The focus in the simulation of morphological instabilities, as in many other moving boundary problems, is the interfacial behavior. It is therefore imperative that a conservative control volume formulation be designed so that the interface velocities and positions can be accurately obtained. Ad hoc discretizations employing finite difference approximations and interpolations at the phase boundary are to be avoided.

4.3   APPLICATION OF THE BOUNDARY-FITTED APPROACH

As reviewed by Crank (1984) and Floryan and Rasmussen (1989), several methods have been proposed in the literature to track moving fronts, each attended by strengths and weaknesses. One commonly used approach involves coordinate transformation to map the irregularly shaped physical domain onto a regular computational domain (Brush and Sekerka, 1989, Lacroix 1989, Shyy 1994, Thompson and Szekely 1989). The numerical methods used in the transformation techniques are well established and are straightforward to implement. The coordinate transformation method is thus a good starting point in developing a field solver and interface tracking ideas. With this in view, in this section we employ a boundary-fitted grid that conforms to the interface and deforms with it.   Since the instability phenomena that determine the microstructure are essentially diffusion-controlled and much of the analytical work in this area has been restricted to this approximation, we only solve the heat conduction equation in each phase. Thus, Eqs. (2.1a) to (2.1d) are reduced to the heat conduction equation in each phase. In addition, the equations representing the effects of surface tension, namely the Gibbs-Thomson condition, Eq. (4.1), and the interfacial velocity given by the Stefan condition, Eq. (4.2), are incorporated in the solution procedure.   The equations are cast in generalized curvilinear coordinates and solved on a transformed domain that is rectangular. From this technique we develop some aspects of the interface tracking scheme to be extended in Chapter 6. In the formulation presented here, unlike mapping methods, it is possible to handle multiple-valued interfaces if a suitable grid can be generated. We also bring to light the limitations of the moving grid formulation in application to the advanced stages of the instability phenomena.

In many moving-grid formulations, the interface motion has been assumed to be slow enough to neglect the terms due to grid motion in the transformed governing equations. This approximation leads to a *quasi-stationary formulation* (Lacroix 1989, Shyy et al. 1993b). We wish to develop an interface tracking procedure that is not restricted to slow moving fronts. Therefore, we first seek to assess the impact of a quasi-stationary approximation on the accuracy of numerical solutions.   In Section 4.4, we reformulate the interface treatment to avoid restriction to nonbranched and isothermal interfaces. The procedure involves straightforward Lagrangian translation of the grid points on the interface. The velocity and geometric parameters (curvature, normal, etc.) at each point are obtained in the physical domain. The proposed method is demonstrated to yield accurate solutions in terms of temperature field and interface location. One also can incorporate, with ease, interfacial phenomena characteristic of smaller length scales such as curvature effects. The computational procedure so developed is then applied to compute the motion of deformed interfaces. The method then will be applied to illustrate the competition among the mechanisms affecting stability, such as surface tension and thermal diffusion.

4.3.1 Formulation

The boundary-fitted grid is first applied to compute the motion of the phase interface for test cases that can be compared with analytical solutions. This will aid in highlighting some aspects of the numerics, particularly accuracy considerations. For a pure conduction problem with phase change between liquid and solid, the governing equations for the energy transfer and interface movement are (Crank 1984), in dimensional form, respectively

$$\frac{\partial T_i}{\partial t} = a_i \left(\frac{\partial^2 T_i}{\partial x^2} + \frac{\partial^2 T_i}{\partial y^2} \right), \quad i = l, s \tag{4.3}$$

$$\varrho L \frac{\partial s}{\partial t} = -k_l \frac{\partial T_l}{\partial n} + k_s \frac{\partial T_s}{\partial n} \tag{4.4}$$

where T_i is the temperature in the ith phase, a_i the thermal diffusion coefficient, and k_i the heat conductivity, L the latent heat of fusion, $x=s(t)$ defines the interfacial position. The partials with respect to n represent derivatives in the direction of the local normal to the interface. Subscripts s and l correspond to solid and liquid, respectively. The second equation is simply the one-dimensional form of the Stefan condition, Eq. (4.2).

The phase front here is macroscopic, and hence the usual macroscopic scales are used. The nondimensional lengths are $X=x/H$, $Y=y/H$, $S=s/H$, H being a representative length scale to be chosen for the specific problem studied. Other nondimensional quantities are $\tau = a_l t/H^2$, $\bar{\varrho} = \varrho/\varrho_l$ and $\Theta = (T - T_m)/(T_\infty - T_m)$, where T_m and T_∞ are the melting temperature and the imposed temperature at the appropriate boundary respectively. It is convenient to express the interface in the form of the function $x=s(y,t)$ (Lacroix 1989). Here we consider heat transport in the liquid phase only. Then, Eqs. (4.3) and (4.4), discarding the subscript, become:

$$\frac{\partial \Theta}{\partial \tau} = \frac{\partial^2 \Theta}{\partial X^2} + \frac{\partial^2 \Theta}{\partial Y^2} \tag{4.5}$$

$$-\frac{\partial \Theta}{\partial n} = \frac{\bar{\varrho}}{St} \frac{\partial S}{\partial \tau} \tag{4.6}$$

where $St = C_p(T_\infty - T_m)/L$ is the Stefan number.

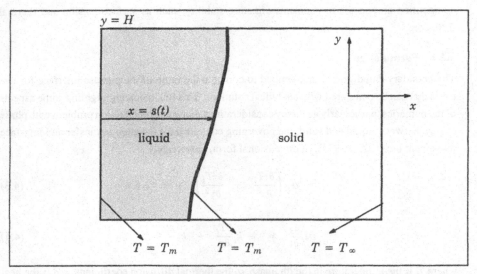

Figure 4.4. *Illustration of the computational domain for comparison with the Neumann solution for a planar phase front*

If $X=S(Y,t)$ represents the interface, then Eq. (4.6), specialized to an isothermal interface, can be rewritten as

$$\left[1 + \left(\frac{\partial S}{\partial Y}\right)^2\right]\frac{\partial \Theta}{\partial X} = -\frac{\tilde{\varrho}}{\text{St}}\frac{\partial S}{\partial \tau} \tag{4.7}$$

Following the transformation from Cartesian (X,Y,t) to curvilinear (ξ,η,τ) coordinates (Thompson et al. 1985) such that $X = X(\xi,\eta,\tau)$, $Y = Y(\xi,\eta,\tau)$, and $t = \tau$, Eqs. (4.5) and (4.6) become:

$$(J\Theta)_\tau + \left(X_\eta Y_\tau - Y_\eta X_\tau\right)\Theta_\xi + \left(Y_\xi X_\tau - X_\xi Y_\tau\right)\Theta_\eta$$
$$= \left[\frac{1}{J}(q_1\Theta_\xi - q_2\Theta_\eta)\right]_\xi + \left[\frac{1}{J}(-q_2\Theta_\xi + q_3\Theta_\eta)\right]_\eta \tag{4.8}$$

$$\frac{1}{J}\left[(Y_\eta\Theta)_\xi - (Y_\xi\Theta)_\eta\right]\left[1 + \left(\frac{1}{J}\left[-(X_\eta S)_\xi + (X_\xi S)_\eta\right]\right)^2\right]$$
$$= -\frac{\varrho}{\text{St}}\left[\frac{\partial S}{\partial \tau} + \frac{1}{J}\left(X_\eta Y_\tau - Y_\eta X_\tau\right)S_\xi + \frac{1}{J}\left(Y_\xi X_\tau - X_\xi Y_\tau\right)S_\eta\right] \tag{4.9}$$

where

$$q_1 = X_\eta^2 + Y_\eta^2 \tag{4.10}$$

$$q_2 = X_\xi X_\eta + Y_\xi Y_\eta \tag{4.11}$$

$$q_3 = X_\xi^2 + Y_\xi^2 \tag{4.12}$$

$$J = X_\xi Y_\eta - X_\eta Y_\xi \tag{4.13}$$

Equations (4.8) and (4.9) constitute the complete set of equations in the generalized coordinates for energy transfer and interface movement. The quasi-stationary approach adopted by many researchers (Lacroix 1989, Thompson and Szekely 1989), simplifies the above equations by dropping all the terms involving coordinate movement, that is X_τ, Y_τ to reduce Eqs. (4.8) and (4.9) to the following forms:

$$\Theta_\tau = \frac{1}{J}\left[\frac{1}{J}\left(q_1\Theta_\xi - q_2\Theta_\eta\right)\right]_\xi + \frac{1}{J}\left[\frac{1}{J}\left(-q_2\Theta_\xi + q_3\Theta_\eta\right)\right]_\eta \tag{4.14}$$

$$\frac{1}{J}\left[\left(Y_\eta\Theta\right)_\xi - \left(Y_\xi\Theta\right)_\eta\right]\left[1 + \left(\frac{1}{J}\left[-\left(X_\eta S\right)_\xi + \left(X_\xi S\right)_\eta\right]\right)^2\right] = -\frac{\varrho}{St}\frac{\partial S}{\partial \tau} \tag{4.15}$$

To test whether such a simplification is acceptable, we design a situation where the interface moves as a planar front. The situation is illustrated in Fig. 4.4. In addition, the temperature in the solid is considered uniform. Thus, the heat transport is effectively one-dimensional and occurs in the liquid phase only. Boundary and initial conditions are given by (Crank 1984),

$$\Theta = 1, X = 0, \tau \geq 0 \tag{4.16}$$

$$\Theta = 0, X \geq S, \tau \geq 0 \tag{4.17}$$

$$\Theta(X, \tau = 0.1) = 1 - \frac{\text{erf}(\frac{X}{2\sqrt{0.1}})}{\text{erf}(\lambda)} \tag{4.18}$$

$$S(X, \tau = 0.1) = 2\lambda\sqrt{0.1} \tag{4.19}$$

where λ is the root of the equation

$$\lambda \exp(\lambda^2) \ \mathrm{erf}(\lambda) \ = \ \frac{\mathrm{St}}{\sqrt{\pi}} \tag{4.20}$$

The exact solution for Eqs. (4.5) and (4.6), with $\tilde{\varrho} = 1$ and with the boundary and initial conditions given in Eqs. (4.16)–(4.19) is

$$\Theta = 1 - \frac{\mathrm{erf}\!\left(\frac{X}{2\sqrt{\tau}}\right)}{\mathrm{erf}(\lambda)} \tag{4.21}$$

The governing equations are discretized in generalized curvilinear coordinates based on a control volume formulation, and a line-SOR procedure is used to solve the flow field. The geometric conservation law discussed in Chapter 2 is to be respected in obtaining the discretized form. Two numerical solution procedures are designed in order to compare the solutions obtained from the complete forms of Eqs. (4.8) and (4.9) and the simplified forms of Eqs. (4.14) and (4.15). These are:

1.   Full treatment: A standard procedure involving backward Euler time-stepping along with a second-order central difference spatial discretization scheme is used to solve Eqs. (4.8) and (4.9). This full set of equations is solved iteratively in a coupled manner to continually update the nonlinear coefficients resulting from the coordinate movement and transformation.

2.   Quasi-stationary treatment: By invoking the quasi-stationary assumption, coordinate movement terms are neglected. On account of Eqs. (4.14) and (4.15), the equations governing energy transfer and interface movement are decoupled and only a simple explicit procedure is needed to update the interface location.

4.3.2   Assessment of the Quasi-stationary Approximation

Three values of the Stefan number were chosen to test the performance of the two numerical approaches above. The values of Stefan numbers and corresponding λ used are based on Eq. (4.20). Eleven grid points are used along the x-direction in each case. For small St, for example, 0.1303 as seen in Fig. 4.5(a), both full and quasi-stationary solutions are in close agreement with the exact solution in terms of the interface position as well as the temperature profile, although the full approach is marginally superior. However, as expected, with increasing St, the quality of solutions obtained from the quasi-stationary method is progressively degraded. In contrast, good agreement is maintained between the solutions of the full approach and the exact solution. Obviously, with a low St, the interface velocity is modest, and the impact of neglecting grid movement terms on the numerical accuracy is not significant. As St becomes larger, however, the simplification to the quasi-stationary approach is no longer acceptable.

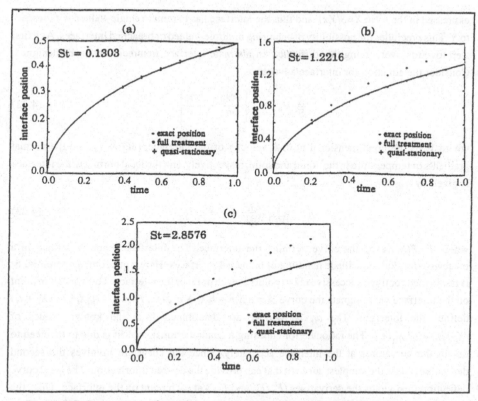

Figure 4.5. Comparison of interface trajectories predicted by the full equations and the quasi-stationary equations with exact solution for three Stefan numbers. (a) St = 0.1303; (b) St=1.2216; (c) St=2.8576.

It may be noted that most of the work on such phase change problems is confined to the low Stefan number regime. Although the Stefan numbers encountered in many naturally occurring processes are small, for some problems of current technological significance, such as the continuous ingot casting of metal alloys (Shyy et al. 1992), Stefan numbers of order 1 are commonly encountered. We have shown that the present solution procedure can handle the interface motion for higher Stefan numbers over periods where the phase boundary has moved to more than twice the original domain size.

4.4 A GENERAL PROCEDURE FOR INTERFACE TRACKING

As mentioned in Section 4.3, Eqs. (4.8) and (4.9) apply only to an isothermal interface. Also, the form of those equations permits calculation of the motion of an interface that can only be

113

expressed in the form $X=S(Y,\tau)$, and thus the interface has to remain single-valued with respect to Y. This precludes the possibility of capturing multiple-valued or branched interfaces. As a first step towards overcoming this difficulty, an alternate interface treatment has to be designed. Consider the equation for interface advance,

$$- \frac{1}{St} V_n = \left[\frac{\partial \Theta_l}{\partial n} - \frac{\partial \Theta_s}{\partial n} \right] \tag{4.22}$$

where V_n is the nondimensional normal velocity of the interface, $(\partial \Theta / \partial n)_{l,s}$ are the normal gradients of temperature in the liquid and solid, respectively, and the local normal to the interface is given by

$$\vec{n} = \frac{1}{|\nabla F|} \left(\frac{\partial F}{\partial X} \vec{i} + \frac{\partial F}{\partial Y} \vec{j} \right) \tag{4.23}$$

where $F=F(X,Y,\tau)$ is the curve defining the interface. The interface shape is defined in a piecewise fashion to facilitate handling of branched interfaces. Here a quadratic polynomial fit is performed for three successive nodal points at each point of the interface. Thus, at the ith point on the interface we designate the curve $Y_i = a_i X_i^2 + b_i X_i + c_i$, i.e., $F_i = Y_i - (a_i X_i^2 + b_i X_i + c_i)$ defines the interface. The a_i, b_i, and c_i are determined from the known values of $(X_j, Y_j), j=i-1,...,i+1$. The rationale for choosing a quadratic functional fit is due to the need to obtain the curvatures at the interface accurately. Since the curvature involves the second derivative, this is the simplest curve fit that can provide the desired information. The local curve definition then yields the derivatives $(F_x, F_y, \text{ and } F_{xx})$ at each point on the interface. Thus, the curvature at each point is obtained from

$$\varkappa = \frac{Y_{XX}}{\left(1 + Y_X^2\right)^{3/2}} \tag{4.24}$$

We may write Eq. (4.22) as

$$- \frac{1}{St} V_n = \frac{1}{|\nabla F|} (F_X \Theta_X + F_Y \Theta_Y) \tag{4.25}$$

In computing the interface normal velocity, then, one seeks to obtain the derivatives F_x, F_y and Θ_x , Θ_y in Eq. (4.25).

The derivatives of temperature may be obtained in the transformed coordinates itself:

$$\Theta_X = \frac{(\Theta Y_\xi)_\eta - (\Theta Y_\eta)_\xi}{J} \tag{4.26}$$

$$\Theta_Y = \frac{(\Theta X_\eta)_\xi - (\Theta X_\xi)_\eta}{J} \tag{4.27}$$

The quantities F_x and F_y, of course, are directly available in the physical domain from the curve fit. Thus, the new coordinates of the interfacial points are obtained from:

$$X^{n+1} = X^n + \frac{\partial F}{\partial X} \frac{v_n}{|\nabla F|} \delta\tau \tag{4.28}$$

$$Y^{n+1} = Y^n + \frac{\partial F}{\partial Y} \frac{v_n}{|\nabla F|} \delta\tau \tag{4.29}$$

where $\delta\tau$ is the time step size. Now that these new coordinates of the curve have been determined, the thermal field is solved for once again, the curve fit is performed at the interface, and a fresh interface position is obtained from Eqs. (4.28) and (4.29). All these procedures are performed in a fully coupled manner involving interaction among the temperature field, interface motion, and grid movement at each iteration. The results for the temperature field in the one-phase, two-phase, and perturbed cases compare well with available analytical results, as will be demonstrated below.

4.4.1 Results and Discussion

4.4.1.1 Case 1. Calculations with Temperature Field Active in One Phase Only.

This alternative interface treatment is compared first with the results of Section 4.3, where the thermal field is active in the liquid region only. The boundary conditions are given by Eq. (4.16–4.19). From Fig. 4.7, it is evident that for all three Stefan numbers tried, namely, 0.1303, 1.2216, and 2.8576, the accuracy of the present scheme is satisfactory and comparable to those in Section 4.3, Fig. 4.5. As can be seen, the computed and exact temperature fields are in excellent agreement. Thus, the current generalized interface tracking procedure does not only cast off the limitations inherent in Eqs. (4.14–4.15) but also yields accurate solutions. In terms of the numerical results, the differences in the interface treatments based on Eq. (4.6) and Eq. (4.25) are significant. It is simpler to directly use Eq. (4.25) in the physical domain, avoiding the undue complexity entailed by the coordinate transformation. Moreover, the latter method is also more generic, capable of incorporating various interface phenomena and addressing more involved interface shapes.

The error introduced in the time-stepping is evaluated for St=0.1303 and St=1.2216. The error value used is the normalized difference in the computed and exact interface locations at nondimensional time $\tau = 0.5$. The error variation for the computed time step sizes depends on the Stefan number and is higher for the computationally more challenging St=1.2216. A Taylor series analysis shows that the time-stepping for the interface as well as temperature fields are first order accurate in δt. However, the global accuracy achieved does not quite reach this order, since in reality the equations for interface motion and temperature field form a spatially and temporally coupled nonlinear system. However, the accuracy is closer to first-order for the lower Stefan number case since the nonlinearity is weaker.

4.4.1.2 Case 2. Calculations with Temperature Field Active in Both Phases

The calculation of interface motion is extended to the case where temperature gradients exist in both solid and liquid. The configuration is as shown in Fig. 4.6. For the case of a planar interface and the domain in Fig. 4.6, the exact solution is given as follows (Crank 1984). The temperature fields in the two phases are given by

$$\frac{\Theta_{out} - \Theta_l}{\Theta_{out} - \Theta_i} = \text{erf}\left[\frac{Y_{out} - Y}{2\sqrt{k_l \tau}}\right] \bigg/ \text{erf}(\lambda) \tag{4.30}$$

$$\frac{\Theta_s - \Theta_o}{\Theta_i - \Theta_o} = \text{erfc}\left(\frac{Y_{out} - Y}{2\sqrt{k_s \tau}}\right) \bigg/ \text{erfc}\left[\lambda \sqrt{\frac{k_l}{k_s}}\right] \tag{4.31}$$

Here, Θ_{out} is the temperature at the outer (liquid) boundary $Y = Y_{out}$, Θ_o is the temperature at the solid boundary $Y = 0$, Θ_i is the interfacial temperature, and $\Theta_{l,s}(Y,\tau)$ are the temperature fields in the liquid and solid phases, respectively.

The interface location is given by

$$S = Y_{out} - 2\lambda \sqrt{k_l \tau} \tag{4.32}$$

and λ is a constant obtained from,

$$\frac{\exp(-\lambda^2)}{\lambda \, \text{erf}(\lambda)} - \frac{\Theta_o - \Theta_i}{\Theta_i - \Theta_{out}} \left(\frac{k_s \varrho_s C_{p_s}}{k_l \varrho_l C_{p_l}}\right)^{1/2} \frac{\exp(-\lambda^2 k_l/k_s)}{\lambda \, \text{erfc}\left(\lambda \sqrt{\frac{k_l}{k_s}}\right)}$$

$$= \frac{1}{\sqrt{\pi} \, C_{pl}(\Theta_i - \Theta_{out})} \tag{4.33}$$

A comparison may be made with the exact solution, with the boundary conditions given as $\Theta = 2$ at the liquid boundary $Y = Y_{out}$, $\Theta = 1$ at the interface, and a time-dependent Θ imposed at the solid boundary $Y = 0$ from the known value of the exact solution. Also, the values St=1 and λ=0.3778 are chosen. The ratios k_l/k_s, c_{pl}/c_{ps}, and ϱ_l/ϱ_s are taken to be 1. The method of solution

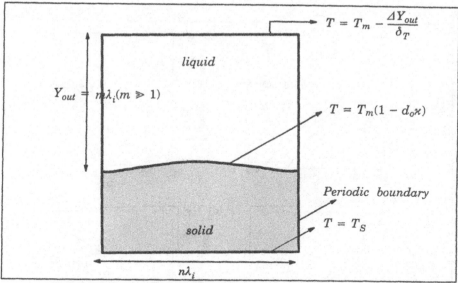

Figure 4.6. Dimensional form of the boundary conditions on the computational domain.

is found to yield good accuracy for the case computed. Good agreement was obtained between the computed and exact temperature fields in the two phases, as shown in Fig. 4.8(a), where the interface location corresponds to $\Theta = 1$. As can be seen from Fig. 4.8(b), the superimposed exact and computed interface locations are practically indistinguishable. The log(error) versus log($\delta\tau$) plot in Fig. 4.8(c) continues to show deviation from the slope expected from the local truncation error analysis, as in the previous case.

4.4.2 Motion of Curved Fronts

4.4.2.1 Interfacial Conditions

The interface tracking method developed in Section 4.4.1 is now applied to follow the development of the phase change front for times that are long enough to obtain substantial curvature of an initially, slightly perturbed interface. In tracking the motion of curved fronts, additional physical and computational issues enter. Physically, when the phase interface is deformed, surface tension seeks to round out regions of strong curvature. This effect is expressed by the Gibbs-Thomson formula,

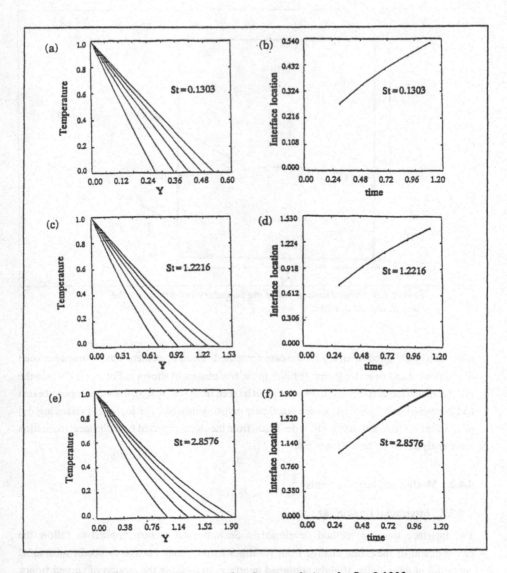

Figure 4.7. Comparison of computed and exact solutions for St=0.1303, St=1.2216 and St=2.8576. The respective Stefan numbers are shown in the boxes. (a), (c), and (e) Exact and computed temperature fields at different time instants. (b), (d), and (f) Superposed exact and computed interface locations versus time.

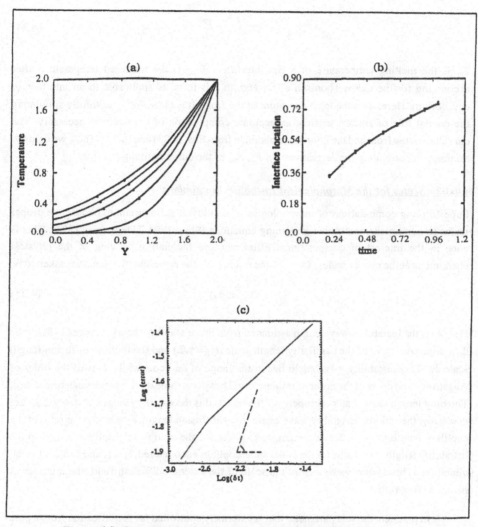

Figure 4.8. Comparison of exact and computed results for the case where temperature gradients are present in both cases. (a) The superposed temperature fields are shown at equal time intervals. As can be seen the solutions are practically indistinguishable. (b) Computed and exact interface positions with time. (c) Error versus $\delta\tau$ plot with stability parameter $\delta\tau/(\delta x)^2$ held constant. Error was obtained at $\tau = 0.5$. The angle drawn represents a slope of 1.

$$T_{ileq} = T_m\left(1 - \frac{\gamma\varkappa}{L}\right) \tag{4.34}$$

T_m is the melting temperature of a flat interface. T_{ileq} is the modified temperature after accounting for the Gibbs-Thomson effect and, in this form, is applicable to an interface in equilibrium. Here, $\varkappa$ is the local curvature of the front. It is clear that in faithfully simulating the crucial role of surface tension, an accurate computation of curvature is necessary. The curvature of the front and the normal are obtained from Eqs. (4.23) and (4.24). Thus, we demand accuracy in computing the derivatives F_x, F_y, F_{xx} of the curve defining the front.

4.4.2.2 Scales for the Morphological Instability Simulations

In performing computations of morphological instability, it is important to choose the proper scales to nondimensionalize the governing equations (Shyy 1994, Udaykumar 1994). This is done in the interest of computational efficiency and facilitating viewing of the physical phenomena at the proper scales. Thus, at the microscale the representative length is taken to be

$$\lambda_c = O(\sqrt{d_o \delta_T}) \tag{4.35}$$

Here λ_c is the instability wavelength estimated from linear stability theory (Langer 1981) to be the geometric mean of the capillarity length scale d_0 (= γ/L) and the thermal diffusion length scale δ_T. This instability wavelength lies in the range of microns, while d_o is of the order of Angstroms and δ_T is of the order of millimeters. Therefore, three distinct scales can be defined. The only length scale that is a property of the material is the capillary length d_o (= γ/L). Let $\varepsilon_1 = \lambda_c/\delta_T$, the ratio of instability wavelength to the diffusion length, $\varepsilon_2 = d_o/\delta_T$, the ratio of the capillary length to the diffusion length, and $\varepsilon_3 = d_o/\lambda_c$, the ratio of the capillary length to the instability length. Typically, for the small undercoolings encountered, ε_1, ε_2, and ε_3 are all small quantities. Thus, when viewed from the scale of the external diffusion field, the interface is planar to first order.

In practice, the sole parameter that is externally controlled for solidification from a pure melt is the melt undercooling, $\Delta = T_l - T_m$, T_l being the temperature imposed at the liquid boundary. The front velocity for a planar front is then a function of Δ and the size of the domain. An appropriate choice of scaling parameters has been found to be critical for computational efficiency. Adopting a temperature scale Δ results in extremely slow computational development of the front due to the resulting small values of nondimensional velocities. Due to the disparity of dimensions, the temperature variations faced by the region adjoining the front are of order $\Delta\lambda_c/\delta_T$ (< $O(\Delta)$) and not Δ.

Thus, with the scales decided upon above, the equation for velocity of the interface becomes

$$\varrho L V_r V_n \;=\; \frac{\varrho a_l C_{pl} \Delta}{\delta_T} \left(-\frac{\partial \Theta_l}{\partial n} + \frac{k_s}{k_l}\frac{\partial \Theta_s}{\partial n}\right) \tag{4.36}$$

Here, V_n is the dimensionless velocity and V_r is the reference velocity scale. For an $\mathcal{O}(1)$ front velocity, we obtain the velocity scale, $V_r = \mathcal{O}(\mathrm{St}\,(a_l/\delta_T))$, where St is again the Stefan number. Equation (4.36) then becomes, in nondimensional form,

$$V_n \;=\; \left(-\frac{\partial \Theta_l}{\partial n} + \frac{k_s}{k_l}\frac{\partial \Theta_s}{\partial n}\right) \tag{4.37}$$

The form of the governing equation adopted depends on the scaling procedure applied. Hence, although Eq. (4.22) and Eq. (4.37) for the interfacial velocity were obtained from the same dimensional form, Eq. (4.2), they differ in appearance.

The time scale of motion of the interface is

$$\tau \;=\; \frac{\lambda_c}{V_r} \;=\; \frac{\lambda_c \delta_T}{a_l \mathrm{St}} \tag{4.38}$$

Hence, the dimensionless form of equation for heat conduction is

$$\varepsilon_1 \,\mathrm{St}\, \Theta_\tau \;=\; \nabla^2 \Theta \tag{4.39}$$

where $\varepsilon_1 = \lambda_c/\delta_T < \mathcal{O}(1)$.

Hence to first-order, the inner temperature field is governed by the Laplace equation and not the unsteady diffusion equation. The diffusion equation is relevant to the interfacial processes only when $\mathrm{St} = \mathcal{O}(1/\varepsilon_1)$, that is when the undercooling is large enough in comparison to L/c_p and the length scale of instability becomes comparable to the diffusion length. However, as the undercooling increases, the length scale of instability decreases.

The equations describing the phenomena at the instability scale now correspond exactly to the macroscopic situation pertaining to viscous fingering in Hele-Shaw cells. This latter issue has been extensively investigated both experimentally and analytically (Pelce 1988, Tabeling et al. 1987, Saffman and Taylor 1958), and there exists the comforting possibility of drawing on the analogous physics thereof. In fact, the only obvious difference between the Hele-Shaw cell and dendritic growth lies in the absence of anisotropy in the former. Such anisotropy has been imposed artificially (Ben-Jacob et al. 1985, Couder et al. 1986), and it turns out that the formerly featureless fingers now develop side-branched structures resembling dendrites.

121

We may now write the following expression for the nondimensionalized interfacial temperature which is obtained from Eq. (4.34):

$$\Theta_{ileq} = \Theta_m \left(1 - \frac{d_o \tilde{\gamma}(\phi)\tilde{\varkappa}}{\lambda_c} \right) = \Theta_m \left(1 - \varepsilon_3 \tilde{\gamma}(\phi)\tilde{\varkappa} \right) = \Theta_m - \gamma_{eff} \, \tilde{\varkappa} \qquad (4.40)$$

Here the reference length scale is λ_c (Eq. 4.35), and the surface tension scale is $L \, d_o$, where L is the latent heat and d_o is the capillary length scale. Also, ε_3 is defined as the ratio of d_o and λ_c, ϕ designates the local interface orientation, and the tilde indicates nondimensional quantities. The curvature has been nondimensionalized in terms of ($1/\lambda_c$). Figure 4.6 illustrates the physical domain on which we carry out our computations and the boundary conditions applied.

4.4.2.3 *Features of the Computational Method*

Equations (4.37–4.40) are solved to follow the motion of a curved interface at the microstructural region. A coupled-implicit treatment along with the generalized interface update procedure is used. A small perturbation is placed on the interface, and a steady-state temperature field is taken as the initial condition. Temperature boundary conditions are applied as described above at $Y=0$, that is, the solid boundary, and at $Y = Y_{out}$, the liquid boundary. At the sides of the domain, we continue to impose an adiabatic boundary condition. Due to the assumed symmetry, the interface is constrained to remain normal to the boundaries at $X = 0$ and $X = X_{out}$. Hence, at the boundaries $F_x = 0$.

An adaptively generated grid distributes grid points in the desired regions. This may be accomplished by spacing the points appropriately along the interface as follows (Shyy 1987, 1994):

Let w_i be a weighting function, corresponding to the ith point along the interface, given by

$$w_i = \left(1 + \frac{\omega \chi_i^2}{\chi_{max}^2} \right) \qquad (4.41)$$

where χ_{max} is the maximum value of the curvature along the interface and ω is an adjustable parameter. The arc length at any i, ψ_i, is obtained from

$$\psi_i = \psi_{tot} \frac{\displaystyle\int_0^\xi \frac{1}{w} d\xi}{\displaystyle\int_0^{\xi_{max}} \frac{1}{w} d\xi} \qquad\qquad (4.42)$$

where ξ is the curvilinear coordinate running along the interface and ψ_{tot} is the total arclength of the front. The integration is performed numerically using the trapezoidal rule.

When $\omega = 0$, the points along the interface are distributed at constant arclengths. For $\omega > 0$, further clustering is achieved in regions of high curvature. It is emphasized in dendritic growth theory that the tip of the dendrite plays a crucial and sensitive role in determining its structure in regard to wavelength and velocity selection. Thus it is desirable to place grid points preferentially at the tip by suitably modifying the constant ω. This procedure is adopted in our calculations below, where ω is assigned the value 0.5 when $\chi_i > 0$ and 0.1 when $\chi_i < 0$.

4.4.3 Results and Discussion

The development of a perturbed interface, on the scale of the morphology, was tracked first for the case of zero surface tension. The initial perturbation of the interface was applied in the form of a cosine wave given by $Y_i = Y_{mean}(1 - 0.05 \cos(2X_i/X_{out}))$. Symmetry was assumed, and hence only half of the perturbation was treated. Forty-one grid points along the X-direction were used. Along Y, 101 grid points in the liquid and 41 points in the solid were used in the calculations.

The width of the computational domain is taken to be 4 (lengths being nondimensionalized by λ_c), which is the half-wavelength of the disturbance. The height Y_{out} is taken to be 50. The interface is originally positioned at $Y_{interface} = 10$. The amplitude of the perturbation is small, being 0.0125 times the wavelength. At the liquid boundary, in accordance with the scaling discussed above, a dimensionless temperature of $\Theta = -40$ is applied, while at the solid boundary at $Y=0$, $\Theta = 1$. Thus, the perturbation faces an undercooled melt, and the Mullins-Sekerka instability (Langer 1981) mechanism is operative. The effect of interface attachment kinetics is neglected in the calculations to focus on the interplay of diffusion and capillarity.

The results for the zero surface tension case in Fig. 4.9 show the rapid development of the instability leading to the splitting of the tip of the protuberance and the formation of a cusp near the boundary. This is consistent with the theory of traveling isothermal fronts or isobaric fronts in the case of Saffman-Taylor fingers (Bensimon 1986, Bensimon and Pelce 1986, Howison et al. 1985). It is found that in such cases, depending on the initial conditions, the disturbance may bifurcate via tip-splitting or may sharpen into cusps. This is because in the absence of surface tension, there is no physical mechanism that provides the short-wave cutoff. The lower limit of resolvable scales is set computationally by the grid spacing. Thus all scales below the single available length scale, namely δ_T, down to the grid spacing, are permissible.

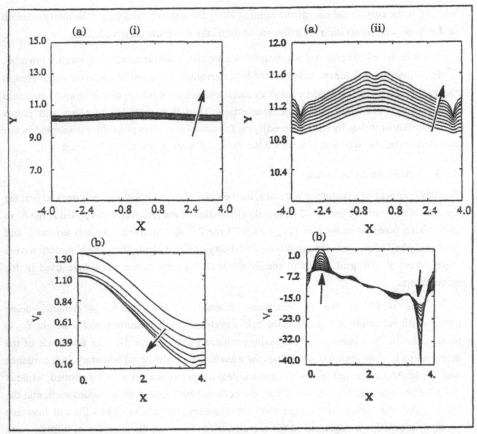

Figure 4.9. Development of a perturbed interface without surface tension. Only the right half was calculated.
(i) Early stages of growth: (a) interface shape after reflection of the half computed; (b) velocity of the right half.
(ii) Later stages of growth: (a) and (b) as above.

Moreover, the sharper the perturbation, the faster it is able to give up latent heat due to the point effect of diffusion, and hence the more unstable it is. This tendency to sharpen can result either in a tip-splitting cascade or in the formation of a cusp. In fact, the appearance of the cusp and the accompanying tip bifurcation ensue quite early in the development of the interface. The initial amplitude to wavelength ratio for the given perturbation, shown in Fig. 4.9(ia) is 0.0125. At the final instant shown in Fig. 4.9(iia), the protuberance has grown to just 8 times its initial amplitude. Shown on the left in Fig. 4.9(ia) is the initial shape of the perturbation. The mean position of the interface at the final stages of the development shown is at $Y=11.2$ for the interface originally positioned at $Y=10$. Thus, the interface does not progress very far along the Y direction before exhibiting tip-splitting and cusp formation. The development of the cusp in the initial stages is seen from the interface velocity plot, shown for the half-wave calculated, in Fig. 4.9(iib). While leading to tip-splitting, from Fig. 4.9(iib), some regions on the interface are travelling faster than their adjacent points. These perturbations will again form distinct protuberances and propagate further until splitting occurs. Again, in the absence of surface tension, there being no lower limit to the length scales allowed, such successive instabilities are physically consistent. Obviously, the interface, corresponding to the discontinuity in the temperature gradient, has not progressed significantly before the instability occurs. The occurrence of the cusp, however, leads to difficulty in further computations, and the calculations are terminated at this point.

The application of surface tension of value $\gamma_{eff} = 0.2$ modifies the interfacial temperature through Eq. (4.40). In this case, the interface development has been computed and lies in strong contrast with results in the absence of surface tension. For such a low value of surface tension, the interfacial temperature assumes values of the order of only $O(10^{-2})$. However, this is sufficient to permit the growth of the perturbation to significant amplitudes without the formation of cusps, as shown in Fig. 4.10(iia). Initially, the entire interface moves upward with speeds that are of the same order. In fact, there is an initial transient, for the given initial conditions when the interface velocity actually decreases. This is shown in Fig. 4.10(ib). However, soon after, the velocity begins to increase again, especially at the tip. In the final stages of the evolution shown in Fig. 4.10(iia), however, the interfacial velocity, shown in Fig. 4.10(iib), is markedly different in two regions. In the region around the tip, the velocity is very much larger than at the sides. This again is due to the differences in temperature gradient between these regions. Regions of negative curvature are warmer at the interface due to the Gibbs-Thomson effect. This is apparent from the temperature contour plots shown in Fig. 4.11, where the isotherms in the liquid are much farther in the region of negative curvature, i.e., at the sides. This implies that this region is warmer than the tip, where the isotherms are highly clustered. This effect, where the interface slows down strongly in the regions near the trough,

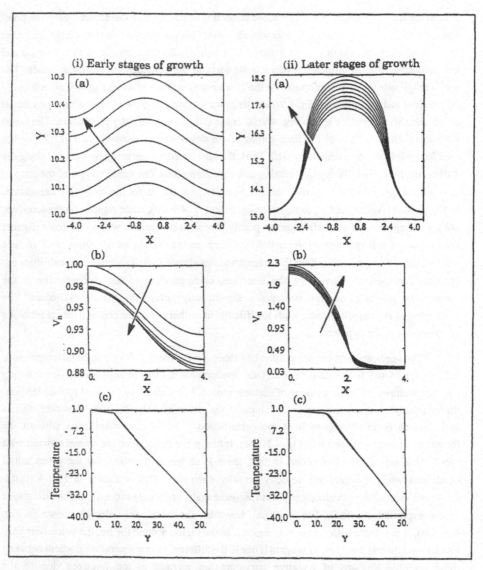

Figure 4.10. Development of perturbed interface with surface tension $\gamma_{eff}=$ 0.2. Arrows show sequence of development. Only the right half was calculated. (i) Early stages of growth: (a) interface shape after reflection of the half computed; (b) velocity of the right half; (c) temperature profile at X=2 in the two phases. (ii) Later stages of growth: (a), (b), and (c) as above.

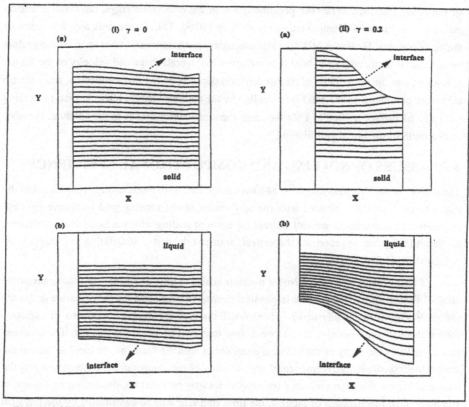

Figure 4.11. Temperature contours in the two phases close to the interface. Contours in the solid are shown only in the range 0 to 0.3. For the case of $\gamma_{eff}=0$, the interface is an isotherm. For $\gamma_{eff} = 0.2$, isotherms cross the interface.
(i) Case without surface tension; (a) Contours in the solid for surface tension $\gamma_{eff}=0$;
(b) Contours in the liquid for $\gamma_{eff}=0$; (ii) Case with surface tension; (a) Contours in the solid for $\gamma_{eff}=0.2$; (b) Contours in the liquid for $\gamma_{eff}=0.2$.

also has been observed by DeGregoria and Schwartz (1986) in their simulations of the Saffman-Taylor fingering phenomenon. This illustrates the common features shared by the analogous situations of solidification and viscous fingering. With surface tension the interface has progressed considerably in contrast to the case without. In fact, the mean interface position now is at $Y=16$ compared to 11.2 in the previous case. The initial amplitude to wavelength ratio of the perturbation is 0.0125 as in the zero surface tension case above. Here the protuberance has grown to nearly 50 times its original size, in contrast with the previous case. This case illustrates the subtle effect that surface tension provides in modifying the stability characteristics of the phase interface. When the perturbation is in the embryonic stage, the surface tension influences the interface temperature only slightly (10^{-2}). Thus it prevents any formation of strong curvatures. However, as the protuberance grows, surface tension has an increasing effect on interfacial temperature and begins to influence the overall shape and velocity of the finger. It is noted that the growth rate of the perturbation depends on the surface tension, since the tip curvature of a bump on the solid is controlled by the capillary length scale. In fact, increasing the value of surface tension to 1 for the cases mentioned above leads to the attenuation rather than growth of the initial perturbation.

4.5   ISSUES OF SCALING AND COMPUTATIONAL EFFICIENCY

The above sections address the physical issues associated with phase change problems and the computational issues associated with the application of the moving grid technique to these problems. In this section, we will address the issue of scaling which is of critical importance in formulating an appropriate theoretical framework and   illustrating its impact on computational efficiency.

Phase change processes involve multiple scales in time and space; thus an appropriate choice of these scales is important. In physical terms, the phenomena at the disparate scales are influenced by different physical mechanisms. Examples are the predominance of capillary forces at small dimensions and the convective transport at large dimensions. Thus, proper scaling enables the study of the physical phenomena with the focus on the mechanisms at the scale under consideration. Numerically, the economy of the computations is determined by the choice of scales. For example, since the time scales may be different, depending on the scales and associated mechanisms of interest, the time step size will be determined by the physical phenomena (e.g., convection, diffusion, capillarity) of significance. In the solidification process, the speed of the interface movement and that of conduction often are not comparable; for low Stefan numbers the interface movement is slow compared to heat conduction. Consequently, if the solidification characteristics are of primary interest, a conduction-based time and length scale may be computationally very inefficient. We shall illustrate this point by presenting a physical example, namely   a practical fiber growth technique using the edge-defined film-fed growth process, schematically illustrated in Fig. 4.12(a). The particular application, discussed in Shyy et al. (1994), is the dynamics of the growth of a sapphire fiber where a crystallographically oriented fibre is grown from a meniscus of molten alumina. The fiber growth system consists of three components: the hot zone, the puller system, and the spooling system. The hot zone consists of an inductively heated refractory metal crucible with edge-defining dies. The melt is supplied to each die tip from a capillary fed manifold. Within the hot zone, the fiber growth process begins with seeding the die tip with an established crystallographic orientation. The pull speed is kept generally constant, and the meniscus dimensions and fiber diameter are manipulated through changes in the induction coil power level setting to adjust the die tip temperature. This type of fiber drawing process is employed in many applications, such as glass products and optical fiber production. Relevant literature can be found in Paek and Runk (1978), Vasilijev et al. (1989), Geyling and Homsy (1990), and Lee and Jaluria (1993).

In the example presented, we concentrate on the transient simulation of the transport and interface dynamics in the hot zone. A static contact condition based on imposed contact angle is imposed on the meniscus which obeys the Young-Laplace equation (Shyy 1994, Shyy et al. 1993a, Shyy et al. 1994) and the governing equations are marched to steady state using the backward Euler scheme. A schematic of the problem domain with the associated boundary conditions is illustrated in Fig. 4.12(b).

4.5.1 Choice of Reference Scales and Resulting Equations

As far as the computational cost is concerned, the choice of the scaling used to nondimensionalize and, more critically, to normalize the various terms in the governing equations, has serious implications on computational efficiency. Two different scaling procedures can be used (Shyy et al. 1994) which result in different forms of the governing equations as follows:

Choice 1: The heat conduction scale is used. The length scale chosen is the radius of the die, r_b; the velocity scale is a_l/r_b; the time scale is r_b^2/a_l; and the temperature scale ΔT is the difference between melting temperature and the die tip temperature. The nondimensional governing equations in axisymmetric coordinates are then written as follows:

Energy equation:

melt:
$$\frac{\partial \theta_l}{\partial \tau} + V_z \frac{\partial \theta_l}{\partial Z} = \frac{1}{R}\frac{\partial}{\partial R}\left(R\frac{\partial \theta_l}{\partial R}\right) + \frac{\partial}{\partial Z}\left(\frac{\partial \theta_l}{\partial Z}\right) \tag{4.43a}$$

solid:
$$\frac{\partial \theta_s}{\partial \tau} + U_p \frac{\partial \theta_s}{\partial Z} = \alpha\left[\frac{1}{R}\frac{\partial}{\partial R}\left(R\frac{\partial \theta_s}{\partial R}\right) + \frac{\partial}{\partial Z}\left(\frac{\partial \theta_s}{\partial Z}\right)\right] \tag{4.43b}$$

Interface movement:

$$\frac{\partial \theta_l}{\partial N} - k\frac{\partial \theta_s}{\partial N} = \frac{\varrho}{\text{St}}\left(U_p - \frac{\partial S}{\partial \tau}\right)_N \tag{4.43c}$$

where U_p is the crystal pull rate and V_z is the advection speed of the melt along the pull direction. V_z is determined by U_p and the reciprocal of the cross-section area of the melt.

Choice 2: The velocity scale is based on the interface movement where the previously defined reference velocity is scaled by the Stefan number. The length scale chosen is the radius of the die, r_b; the velocity scale is St α_l/r_b; the time scale is St r_b^2/α_l; and the temperature scale ΔT is the difference between the melting temperature and the die tip. The governing equations then become:

Energy equation:

melt:
$$\text{St} \left[\frac{\partial \theta_l}{\partial \tau} + V_z \frac{\partial \theta_l}{\partial Z} \right] = \frac{1}{R} \frac{\partial}{\partial R} \left(R \frac{\partial \theta_l}{\partial R} \right) + \frac{\partial}{\partial Z} \left(\frac{\partial \theta_l}{\partial Z} \right) \tag{4.44a}$$

solid:
$$\text{St} \left[\frac{\partial \theta_s}{\partial \tau} + U_p \frac{\partial \theta_s}{\partial Z} \right] = \alpha \left[\frac{1}{R} \frac{\partial}{\partial R} \left(R \frac{\partial \theta_s}{\partial R} \right) + \frac{\partial}{\partial Z} \left(\frac{\partial \theta_s}{\partial Z} \right) \right] \tag{4.44b}$$

Interface movement:

$$\frac{\partial \theta_l}{\partial N} - k \frac{\partial \theta_s}{\partial N} = \varrho \left(U_p - \frac{\partial S}{\partial \tau} \right)_N \tag{4.44c}$$

Figure 4.13 shows the relative computing efficiency of the two scaling procedures for low Stefan numbers, that is, Stefan numbers less than unity. In each case the final solution was the steady-state temperature field in the melt and crystal as well as the shape assumed by the meniscus at steady state. Shown in the figure is the time variation of the height and radius of the crystal. Substantially different convergence behavior can be observed between the two scaling procedures. With Choice 1, the interface speed is of $O(\text{St})$, while with Choice 2 the interface speed is of $O(1)$. Consequently, Choice 2 needs far fewer number of time steps to reach the steady state solution. Furthermore, this illustrates that with Choice 2, for low St the energy equation becomes quasi-stationary due to the fact that that the thermal field relaxes almost instantaneously to accommodate the interface movement. Hence the computing costs of the two procedures differ by $O(\text{St}^{-1})$, making Choice 2 appropriate for low St computations. The key concept here is that the reference scales should be chosen so that the nondimensional interface speed is $O(1)$. Further details can be found in Shyy et al. (1994).

4.6   CONCLUSIONS

In this chapter, some salient features of moving boundary problems have been addressed in the context of solidification processes. The treatment of the *whole* physical system, including the moving boundary, has been emphasized. At a very fundamental level, the main difficulties pointed out by Marchaj (1979) concerning sail design can be generalized to a encompass a

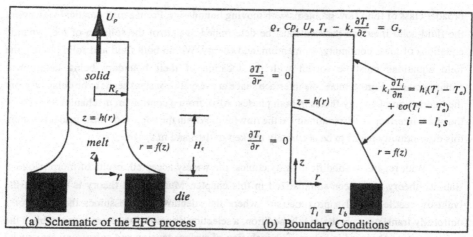

Figure 4.12. *Illustration of the edge-defined film-fed growth (EFG) technique and the associated heat transfer mechanisms.*

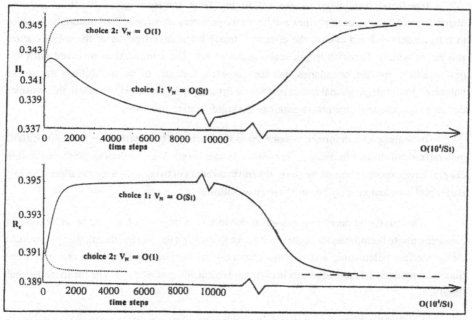

Figure 4.13. *Effect of scaling procedure on the convergence characteristics. St = 0.024. Choice 1: $V_n = O(St)$, and Choice 2: $V_n = O(1)$.*

broader class of fluid flow problems with moving boundaries. From a mathematical viewpoint, the fluid-solid interface position must be determined as part of the solution of the dynamic equations of mass continuity, momentum, and energy. Within both fluid and solid phases, the field equations must be solved with the location of their boundary being determined simultaneously. Furthermore, the interface shape at every time instant needs to be obtained from the interaction of energy fluxes in both phases. Also, from a continuum mechanics viewpoint, the phase interface is a discontinuity in the flowfield. Within the context of finite grid resolution, this discontinuity needs to be accurately tracked in time and in space.

With regard to solidification dynamics, the widely accepted results of morphological stability theory have been summarized in this chapter. Much of the theory is based on the Ivantsov needle crystal approximations, where the dendrite stem assumes the shape of a uniformly translating paraboloid. In addition, a selection principle is necessary to establish the unique pattern observed in nature. The inclusion of surface tension and anisotropy are crucial to the emergence of selection from the governing equations.

Interfacial instabilities at the solidifying front influence grain structure and the distribution of dopants or impurities and hence the performance of solidified materials. In order to fully understand and control the effects of instabilities, investigation of the solidification process at widely disparate length scales is necessary. The complexities involved with the nonlinearities, moving boundaries, and the microscopic nature of the instabilities has so far hampered both computational and experimental investigation. An effort is made in this work to devise a computational method to simulate the solidification processes.

In dealing with deformed boundaries with structured grid methods, a boundary-fitted, generalized curvilinear formulation is commonly employed. The simulations presented in this chapter have exposed some of the strengths and drawbacks of using such a formulation. The key issues and conclusions may be summarized as follows.

1.   The quasi-stationary approximation for interface motion is found to be inaccurate in capturing phase boundaries for higher Stefan numbers. With the full treatment, the temperature field, interface calculation, and grid movement all interact over each iteration and time step. Such a computational method results in close agreement with analytical results for an isothermal interface.

2.   A generalized interface motion technique has been successfully developed, which is not limited to a single-valued or isothermal interface. Computations show that this method produces results in close agreement with analytical solutions for the one-phase and two-phase cases.

3.   The spatio-temporal development of deformed interfaces, with the incorporation of the Gibbs-Thomson effect due to surface tension, is in accordance with theory. The subtle effect of surface tension in preventing formation of cusps and stabilizing the interface are borne out in the results. Tip-splitting and cusp formation are observed in the absence of surface tension. Addition of a small amount of surface tension completely changes the picture, and a propagating finger is obtained for the duration of the computation. Thus, unless such features of the interface as curvature are computed accurately, and the temperature boundary condition is imposed carefully in the course of the computation, this sensitivity with respect to surface tension may be lost. The shape and behavior of the finger are in qualitative agreement with other investigators of the analogous viscous fingering phenomenon. In conjunction with the scaling procedure adopted, significant morphological development has been achieved in a computationally efficient manner. The results obtained clearly show the effects of the competing mechanisms on the stability of a perturbed interface.

4.   Although the interface tracking method developed here can handle a multiple-valued situation, there are several caveats attached to a procedure designed on the framework of this purely Lagrangian approach. Firstly, as the interface dilates, the markers accumulate in regions where the interface is compressed and rarify in regions that are stretched. Eventually, this undesirable nonuniform distribution of particles can lead to deterioration of the interface information. Thus, in order for Lagrangian tracking to be viable, the interfacial markers will need to be periodically rearranged along with the grids in the vicinity of the interface. Also, it is not clear how the method could be extended in the event of topological changes of the interface. In spite of these limitations, the method can be useful in elucidating various aspects of the phase change process as long as the interface is not too deformed.

Due to the limitations imposed by the need for the grid to conform to the moving interface, the boundary-fitted formulation is unattractive as a simulation technique for highly distorted fronts. Thus, if a technique can be designed to disconnect grid motion from that of the interface so that the interface is the only moving component, greater flexibility in dealing with such situations could be achieved. We will pursue this objective later, in Chapter 6. In the next chapter, a fixed grid technique based on the enthalpy formulation is presented to solve a macroscopic phase change problem. From the contrasting requirements and merits of the moving grid and enthalpy formulations, the need for the combined Eulerian-Lagrangian approach developed in later chapters will become clear.

CHAPTER 5

FIXED GRID TECHNIQUES: ENTHALPY FORMULATION

In Chapter 4, phase change dynamics with conductive transport has been discussed. Here, we present a fixed-grid formulation for the phase change problem including convective transport. First, the governing equations of fluid flow, heat transfer, and the jump conditions at the phase change interface are presented. A scaling analysis at the transport and morphological scales is then presented to illustrate the wide disparity of the length and time scales that are present in a typical solidification system. We then cast the equations in a space-averaged form (enthalpy formulation) and illustrate its application to a practical crystal-growth problem.

5.1   GOVERNING EQUATIONS

The following equations are solved in the liquid phase:

Continuity:
$$\nabla \cdot \vec{u} = 0 \tag{5.1}$$

Momentum:
$$\frac{\partial \vec{u}}{\partial t} + \vec{u} \cdot \nabla \vec{u} = \nu \, \nabla^2 u - \frac{\nabla p}{\varrho_o} - g \, \vec{j} \tag{5.2}$$

Energy:
$$\frac{\partial T}{\partial t} + \vec{u} \cdot \nabla T = a_l \nabla^2 T \tag{5.3}$$

In the solid phase:
$$\vec{u} = 0 \tag{5.4}$$

Energy:
$$\frac{\partial T}{\partial t} = a_s \nabla^2 T \tag{5.5}$$

where, for simplicity, the transport properties are held constant. The equations in each phase are solved separately subject to the boundary conditions on the interface:

Gibbs-Thomson condition: $\quad T_{interface} = T_m \left(1 - \frac{\gamma}{L} \varkappa\right)$ $\qquad$ (5.6)

where T_m is the melting temperature, γ is the surface tension, L is the latent heat of fusion and $\varkappa$ is the local curvature of the interface. The fluid velocity boundary conditions are:

no-slip: $\qquad\qquad\qquad \vec{u} \cdot \vec{t} = 0$ $\qquad$ (5.7)

mass conservation: $\qquad \varrho_L \vec{u} \cdot \vec{n} = (\varrho_s - \varrho_L)\, V_N$ $\qquad$ (5.8)

where V_N is the normal velocity of the interface, given by,

Stefan condition: $\qquad \varrho_s L V_N = \left(k_s \frac{\partial T_s}{\partial n} - k_l \frac{\partial T_l}{\partial n}\right)$ $\qquad$ (5.9)

Here subscripts l and s stand for liquid and solid, respectively. Appropriate Neumann and Dirichlet conditions are applied at the boundaries of the computational domain.

5.2 SCALING ISSUES

As already discussed in Chapter 4, a striking difficulty in analyzing the solidification and melting problem is the presence of many dimensionless parameters, frequently numbering ten or more. One of the basic problems encountered by any theoretical model is the wide disparity of length scales. Based on the dimensional analysis, several length scales can be identified in a binary eutectic system, namely, (i) capillary length d; (ii) thermal transport length scale δ_T; (iii) solutal transport length scale δ_s; and (iv) convective length scale δ_v. There is more than one way to choose a convective length scale, depending on the processing technique as well as the operating conditions. For example, in Bridgman growth, the convective length scale can be determined by balancing buoyancy and shear stress effects, while in a floating zone technique it can be deduced from the balance of Marangoni and shear stress effects. The overall solidification process is very complicated, involving the simultaneous presence and competition of multiple scales. For the metallic system, due to the disparity of Schmidt and Prandtl numbers ($Sc \gg 1$, $Pr \ll 1$), the following relative order of magnitudes can be identified among the length scales (Shyy 1994):

$$O(d) \ll O(\delta_s) < O(\delta_v) < O(\delta_T) \qquad (5.10)$$

The appropriate magnitude of the morphological length scale, λ_c, e.g., the dendritic spacing, depends on the balance resulting from these different scales and associated transport mechanisms. It should be noted that λ_c need not be the same as any of the above length scales. Its order of magnitude can be different from those already established by the capillary and transport mechanisms. The reason that this extra morphological scale can exist is due to the nonlinearity of the system. A case in point is the well known morphological stability analysis performed first by Mullins and Sekerka (1964), which yields a relationship of $\lambda_c = O\left(\sqrt{d\delta_T}\right)$ for an interface subject to directional solidification, where d ($= \gamma/L$) is the capillary length scale. However, this analysis does not consider the influence of convection. An analytical approach must address the difficulties involved with the multiple scales along with the nonlinearity of the system. In order to fully account for the various length scales that coexist in the melt, a direct numerical simulation that accounts for the interaction among different length scales is perhaps the only viable approach.

We now present an example of a scaling procedure that enables information transfer from the macroscopic to microscopic scales. The scaling performed here is restricted to the particular configuration chosen below and to a pure material. In the case of impure materials/alloys the scaling is considerably more complicated due to the addition of other competing effects and control parameters.

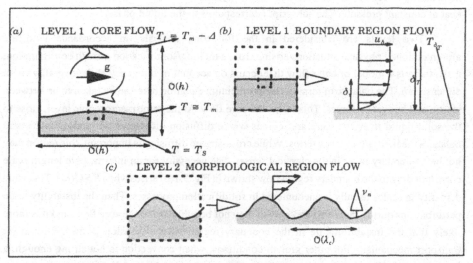

Figure 5.1. Illustration of the levels involved in analysis of convective effects on microstructure evolution.
(a) Level 1 – core flow (b) Level 1 – flow near the macroscopic interface
(c) Level 2 – flow at the morphological level

Let us assume the growth configuration shown in Fig. 5.1 for the solidification/melting of a pure material. Let the domain size be defined by the length h. Let, as in the experiments of Glicksman et al. (1976), a global undercooling, Δ, be imposed and T_m be the melting temperature. The flow is assumed to be purely buoyancy-induced. Sources of forced convection, although inherent in any realistic solidification process, are ignored. Estimates for the scales at various levels will be obtained based on these imposed conditions and the material properties. The following nondimensional quantities can be defined,

$$Pr = \text{Prandtl number} = \nu_l/a_l \tag{5.11}$$

$$Ra = \text{Rayleigh number} = \frac{g\,\beta_l\,\Delta\,h^3}{a_l\,\nu_l} \tag{5.12}$$

$$Gr = \text{Grashof number} = Ra \cdot Pr \tag{5.13}$$

$$St = \text{Stefan number} = \frac{C_p\Delta}{L} \tag{5.14}$$

where Δ, h, β, a_l, ν_l, L, and C_p are respectively the global undercooling, global length scale, thermal expansion coefficient, thermal diffusivity, kinematic viscosity, latent heat, and specific heat at constant pressure. The subscript l corresponds to the liquid phase.

Two levels are considered in the following analysis, in accordance with the aforementioned physical situations arising from a melting/solidification growth configuration. First, there is a global or outer flow that occurs on account of buoyancy at the overall growth dimension h. The balance of terms in the momentum equation here can be taken to be between inertia and buoyancy forces. This forms the core flow. At the same macroscopic level, close to the solid-liquid interface there are regions where diffusion of momentum predominates and balances the inertia/buoyancy terms. While cross-stream diffusion is important, this region may not be a boundary layer in the classical sense. The necessity for an intermediate length scale corresponding to the boundary region flow shown in Fig. 5.1(b) arises when $h > O(\lambda_c)$. This wide disparity in scales usually is encountered in solidification processes. Then the instability scale perturbations during inception and growth may not be subjected to the core flow, and it is more likely that the instabilities lie in the boundary region where diffusion is the predominant transport mechanism. For other growth conditions, when convection is not strong enough, a distinct intermediate layer may not exist. In that case, the matching of the microscopic or inner region can be performed directly with the core flow. The morphological scales represent the next level, as shown in Fig. 5.1(c). The scales at this level are determined from pattern formation theory, which assumes a diffusion dominated situation. The critical feature at this scale is the interaction of heat conduction, latent heat release at the moving boundary, and capillarity effects.

To determine the effects of fluid flow at the global level on the microstructure, it is necessary to establish communication between the two levels both with and without an intermediate layer, i.e., the boundary region. A sample scale analysis is presented in the following sections.

5.2.1 The Macroscopic Scales

For the outer flow field, such as the core flow in Fig. 5.1(a), the length scale is taken to be h and the temperature scale is taken as $(T_l - T_m)$. The subscript g indicates a global scale. T_l is the imposed constant temperature at the boundary of the liquid phase. The velocity scale along x-direction, U_g, is obtained by considering the balance of transport terms in the continuity and momentum equations. Continuity indicates that the y-direction flow velocity in the melt, V_g, is $O(U_g)$. Considering the momentum equations, and eliminating pressure by cross-differentiation, we obtain for balance between convection and buoyancy terms a buoyancy velocity scale,

$$U_g = \sqrt{g\beta_l \Delta h} \tag{5.15}$$

Thus, the characteristic time scale is

$$\tau_g = \frac{U_g}{h} = \sqrt{\frac{g\beta_l \Delta}{h}} \tag{5.16}$$

The velocity scale of interface motion as viewed at the outer scale is

$$V_{n_g} = \frac{k_s \Delta}{\varrho_l h} = \frac{\Delta}{L/C_p}\frac{a_l}{h} = \frac{a_l \mathrm{St}}{h} \tag{5.17}$$

This interfacial velocity scale is based on the macroscopic perspective in which the interface is only mildly deformed. The interface then is driven by the gradients of the temperature field, and the diffusion of latent heat determines the extent of the temperature boundary layer and hence the front velocity. It is noted that at the level of the microstructure, a selection principle is employed to estimate the front velocity. At that scale the application of the selection principle incorporates the crucial effects of capillarity and anisotropy, aspects which can be safely ignored at the macroscopic level. The ratio of the front velocity to the buoyancy velocity scale is obtained from

$$\frac{V_{n_g}}{U_g} = \frac{\mathrm{St}\,h}{\delta_T \sqrt{\mathrm{Pr}\,\mathrm{Ra}}} \tag{5.18}$$

For the small St and large Ra commonly encountered in solidification processing Eq. (5.18) indicates that the interfacial velocity scale is such that $V_{n_g} < O(U_g)$. Thus, at the outer scales the quasi-stationary assumption is reasonable. The interface is thereby assumed to move slowly enough that its motion can be neglected in comparison to the fluid flow velocities.

As a next step we require scales in the region near the solidifying/ melting interface. If such a region exists, the physical mechanisms that balance are diffusion and inertia/buoyancy. The matching between the core flow and the microscopic scales will have to be performed through this region. Hence, this boundary region will form an intermediate layer between levels I and II. In this boundary region estimates are required for:

1. δ_v, the intermediate viscous scale
2. δ_T, the intermediate thermal length scale
3. U_{δ_T}, the velocity scale in the thermal layer
4. U_{δ_v}, the velocity scale in viscous layer
5. Δ_δ, the temperature scale in the intermediate layers, appropriately subscripted to indicate the length scale in question.
6. τ_δ, the time scale in the intermediate layer

These quantities are to be obtained in terms of the global quantities Δ, h and material properties, or alternatively in terms of the non-dimensional numbers in Eq. (5.11–5.14).

Although, as mentioned before, the flows in the intermediate layers may not necessarily correspond to boundary layers but merely represent a region where diffusion plays a dominant role, some features of boundary layers need to be noted. In the classical boundary layer theory, depending on the Prandtl number very different scenarios are possible, as illustrated in Fig. 5.2. Such a distinction is important to make; since the Prandtl numbers for metallic melts and the organic melts used to model their behavior are vastly different, care must be exercised in extrapolating results from experiments involving organic melts to metals, particularly when convective effects are involved. Here we obtain estimates for the above scales for each case.

Consider a typical solidification arrangement shown in Fig. 5.3 where the core flow is driven by buoyancy. Different types of flow situations can be encountered depending on the global flow characteristics (Davis 1990, Favier 1990, Ostrach 1983). These situations are illustrated in Fig. 5.3. In type 1, there is a corner flow situation, and the microstructures formed in this region will be strongly influenced by wall effects. In type 2, a stagnation point flow arises between counter-rotating vortices in the bulk flow. In type 3, a boundary layer over the curved macroscopic interface is encountered for $Gr > O(1)$. Thus, several estimates of the flow characteristics and intermediate scales for the flow field are relevant. It is clear that in any

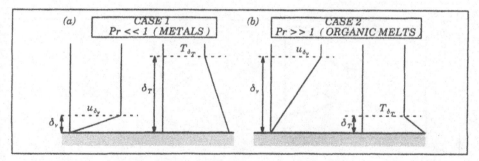

Figure 5.2. Illustration of the boundary layer types for different Prandtl numbers (a) Case 1, Metallic melts. (b) Case 2, Organic melts used to model the former.

realistic arrangement, different regions of the macroscopic interface and hence the microstructures there experience a variety of transport phenomena. In addition, depending on the forces driving fluid motion, the interface can experience different conditions from onset to fully developed flow (Davis 1990, Davis et al. 1984, Favier and Rouzaud 1983, Forth and Wheeler 1989, McFadden et al. 1984). In what we are presenting we limit ourselves to the situation in type 2 shown in Fig. 5.3(d) in order to investigate the coupling of the global and microscopic level transport processes in their simplest form.

5.2.2 Velocity Scales

Applying the usual boundary layer considerations, we can write the scales in the viscous layer as

$$X_{\delta_v} = h \tag{5.19}$$

and
$$Y_{\delta_v} = \delta_v \tag{5.20}$$

where the subscripts represent the scales in the intermediate viscous layer of extent δ_v. The continuity equation yields, for the melt velocity along the y-direction

$$V_{\delta_v} = \frac{\delta_v U_{\delta_v}}{h} \tag{5.21}$$

This assumption is predicated on the condition that $V_{n_{\delta_v}} < O\left(V_{\delta_v}\right)$, that is, that the velocity of the interface, as viewed from within the intermediate layer, is negligible compared to the v-component of the flow velocity there. The validity of this assumption depends on the ratio between the interface velocity as viewed from the viscous layer $V_{n_{\delta_v}}$ and the fluid flow velocity in that region given by Eq. (5.21):

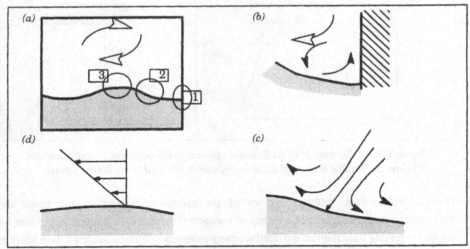

Figure 5.3. *Flow features in a typical solidification arrangement.*
(a) The macroscopic interface. (b) type 1– corner flow (c) type 2– stagnation point
flow (d) type 3 – boundary layer flow along a curved surface.

$$\frac{V_{n_{\delta_v}}}{V_{\delta_v}} = \frac{(St\,a_l/\delta_T)}{(\delta_v\,U_{\delta_v}/h)} = \frac{St\,a_l}{\delta_T}\frac{\sqrt{Re}}{\sqrt{g\beta_l\Delta h}} \tag{5.22}$$

where δ_T is the intermediate thermal length scale, $Re = \sqrt{Gr}$ and

$$V_{n_{\delta_v}} = \frac{St\,a_l}{\delta_T} \tag{5.23}$$

since the interface is driven by gradients of the temperature field and this field relaxes over an extent δ_T. This point will be addressed in more detail later. Therefore, as far as the interface motion is concerned, the length scale used for computing the temperature gradients should be the thermal layer thickness δ_T. The second equality in Eq. (5.22) results from the well-known fact that the viscous layer follows the scale

$$\delta_v = \frac{h}{\sqrt{Re}} \tag{5.24}$$

which by the heat transfer analogy $Gr = Re^2$ can be written as

$$\frac{\delta_v}{h} = \left(\frac{Pr}{Ra}\right)^{\frac{1}{4}} \tag{5.25}$$

This can be simplified to the result

$$\frac{V_{n_{\delta_v}}}{V_{\delta_v}} = \frac{St\,h}{\delta_T} \tag{5.26}$$

Thus the consistency of the assumption $V_{n_{\delta_v}} < O\left(V_{\delta_v}\right)$ can be judged once the magnitude of δ_T is estimated.

5.2.3 Thermal Scales

5.2.3.1 Low Prandtl Number (Metallic Melts)

The following condition applies

$$Pr < O(1) \tag{5.27}$$

which implies that

$$\delta_T > O(\delta_v) \tag{5.28}$$

Hence from Fig. 5.2 we estimate that

$$U_{\delta_v} = U_g = U_{\delta_T} \tag{5.29}$$

i.e., in the bulk of the temperature boundary layer the velocity scale is $U_g = U_{\delta_v}$.

From the situation shown in Fig. 5.2(a), this situation applies to metallic melts and

$$Pr < O(1) \quad , \quad \delta_v < O(\delta_T) \tag{5.30}$$

For this case, from Fig. 5.2(a) we can assume the following scales for the velocity:

$$U_{\delta_T} = U_{\delta_v} = U_g = \sqrt{g\beta_l\varDelta h} \tag{5.31}$$

For convection-diffusion balance in the energy equation,

$$\frac{U_{\delta_T}\varDelta}{h} \sim \frac{a_l\varDelta}{\delta_T^2} \tag{5.32}$$

$$\Rightarrow \delta_T = \sqrt{\frac{a_l h}{U_{\delta_T}}} \tag{5.33}$$

which, upon some manipulation, yields

$$\delta_T = O(h\,Pr^{-1/2}\,Re^{-1/2}) = O(h\,Pr^{-1/4}\,Ra^{-1/4}) \tag{5.34}$$

In this thermal boundary layer of thickness δ_T, then

$$U_{\delta_T} = U_g = \sqrt{g\beta_l\varDelta h} \tag{5.35}$$

143

$$V_{\delta_T} = O\left(\frac{\delta_T}{h}\sqrt{g\beta_l \Delta h}\right) \qquad (5.36)$$

$$\Delta_{\delta_T} = \Delta \qquad (5.37)$$

Now, we return to the ratio in Eq. (5.22) in order to determine the range of validity of the scales above. Thus,

$$\frac{V_{n_{\delta_T}}}{V_{\delta_T}} = \frac{St}{Pr^{-1/2}Re^{-1/2}} = St\,Pr^{1/4}\,Ra^{1/4} = St\,Gr^{1/4} \qquad (5.38)$$

From this relationship, it is evident that if

$$Gr^{1/4}\,St > O(1) \qquad (5.39)$$

that is,.,

$$Gr > O\left(St^{-4}\right) \qquad (5.40)$$

then the interfacial velocity magnitude will become comparable to V_{δ_T}, and thus the boundary layer picture will change. Therefore, the above scales are valid only under the conditions

$$Pr < O(1) < Gr < O\left(St^{-4}\right) \qquad (5.41)$$

In most cases of interest the Stefan numbers are small. Thus there is a range of values of the Grashof number under which the above conditions hold. The first inequality is necessary for a boundary layer type situation to exist, while the second inequality places an upper bound on the Grashof number based on the interface velocity so that a *slow-growth* condition can be applied. This latter condition permits the morphological stability theory so that the scales at the microscopic level can be obtained. However, for some practical applications such as rapid solidification processing (Kurz and Trivedi 1992), this approximation is not valid.

5.2.3.2 High Prandtl Number (Organic Melts)

The scales are obtained as follows

$$Pr > O(1) \qquad (5.42)$$

$$\delta_T < O(\delta_\nu) \qquad (5.43)$$

Hence the appropriate velocity scale in the thermal boundary layer is given by

$$U_{\delta_T} = \frac{\delta_T U_{\delta_\nu}}{\delta_\nu} = \frac{\delta_T U_g}{\delta_\nu} \qquad (5.44)$$

As seen in Fig. 5.2(b), for this case

$$Pr > O(1), \delta_\nu > O(\delta_T) \qquad (5.45)$$

The order of magnitude of velocity in the layer of thickness δ_T , designated U_{δ_T}, is given by

$$U_{\delta_T} = \frac{\delta_T U_g}{\delta_v} \tag{5.46}$$

From continuity, again assuming that $V_{\delta_T} > V_{n\delta_T}$,

$$V_{\delta_T} = \frac{U_{\delta_T} \delta_T}{h} = \frac{\delta_T^2 U_g}{\delta_v} \tag{5.47}$$

For convective-diffusive balance in the energy equation,

$$\frac{U_{\delta_T} \Delta_\delta}{h} \sim \frac{a_l \Delta_\delta}{\delta_T^2} \tag{5.48}$$

which after simplification implies that

$$\frac{\delta_T}{h} = O(Pr^{-1/3} Re^{-1/2}) = O(Ra^{-1/4} Pr^{-1/2}) \tag{5.49}$$

In this case, from the continuity equation

$$V_{\delta_T} = \frac{\delta_T U_{\delta_T}}{h} \tag{5.50}$$

and again, due to the same arguments as for Eq. (5.23) above,

$$V_{n_{\delta_T}} = O\left(\frac{St\, a_l}{\delta_T}\right) \tag{5.51}$$

Therefore the scaling is valid only if $V_{n_{\delta_T}} < V_{\delta_T}$, and this condition is now represented by the ratio

$$\frac{V_{n_{\delta_T}}}{V_{\delta_T}} = St\, Re^{1/2} = St\, Ra^{1/4} Pr^{-1/4} \tag{5.52}$$

Therefore, the scaling breaks down if

$$Ra^{1/4} Pr^{-1/4} St > O(1) \tag{5.53}$$

$$\Rightarrow \frac{Gr^{1/4}}{Pr^{1/2}} St > O(1) \tag{5.54}$$

that is, if

$$Gr > O\left(\frac{Pr^2}{St^4}\right) \tag{5.55}$$

Since in this case $Pr > O(1)$ and $St < O(1)$ for most solidification processes, the scaling will hold for much larger Grashof numbers than in case 1.

5.2.4 The Morphological Scales

At the morphological level, several different scalings were performed by various authors (Brush and Sekerka 1989, Langer 1980). The difficulty in choosing the scales of the microstructural evolution stems from the classic difficulty of separating the scales of length and velocity, that is, in determining a selection criterion. As mentioned before several characteristic lengths and times exist for the interfacial features at the microscopic levels. The scales that one adopts therefore depend on the features that one is interested in capturing. In the course of our work, it became apparent that an injudicious choice of scales will lead to inordinate amount of computational effort. This was due to conflicting requirements of characteristic physical times and time-stepping criteria imposed by numerical stability considerations. For instance, in the diffusion controlled phenomena, a natural choice for the time scale would be a diffusion time scale. However, it is well known that for low Stefan numbers, the interface motion is slow enough that the diffusion field relaxes almost instantaneously to transport the latent heat generated at the moving interface. Thus the quasistationary behavior of the field renders the diffusion time scale inappropriate in tracking the evolution of the phase front. A better choice would be the time scale of motion of the front. Similar considerations apply to the other aspects and variables in interface tracking at the microscopic level. In the following we present a scaling procedure for the microscale.

In analyses of morphological instabilities, it is common practice to adopt the instability wavelength from linear stability considerations as the length scale (Langer 1980). As already discussed, this scale is given by

$$\lambda_c = \sqrt{d \delta_T} \tag{5.56}$$

In diffusion controlled growth it is not possible to explicitly determine the length λ_c. This is because the thermal boundary layer thickness, $\delta_T = \alpha_l / V_n$, cannot be estimated theoretically for a general growth configuration. The growth velocity can be controlled in practice through the imposed thermal conditions. The instability length remains dependent on the front velocity, a classical problem of selection. The quantity δ_T therefore is estimated to obtain λ_c. A capillary length scale can also be defined as

$$d' = \frac{\gamma C_p T_m}{L^2} \tag{5.57}$$

The velocity scale is chosen so that the nondimensional interface velocity is O(1) and to facilitate efficient computations and to focus on the important physical mechanisms. Here it is important to adopt an appropriate inner temperature scale. The situation is shown in Fig. 5.4. It is clear that an appropriate inner temperature scale is

$$\Delta_l = \frac{\Delta \lambda_c}{\delta_T} \tag{5.58}$$

Substituting in the Stefan condition Eq. (5.9) gives

$$\varrho_l L V_n v_n = \frac{\varrho_l a_l C_p \Delta}{\delta_T} \left[\frac{\partial T^*}{\partial n^*} \right]_{s-l} \cdot \vec{n} \tag{5.59}$$

where v_n is the nondimensional normal front velocity, and thus the front velocity is

$$V_n = \mathcal{O} \left(\frac{a_l \text{St}}{\delta_T} \right) \tag{5.60}$$

The time scale has to be chosen carefully. A straightforward time scale choice would be the diffusion scale, but this may be inappropriate for observing interfacial processes. The time scale of the interface evolution is a slow scale, while that of diffusion is a fast scale. For tracking the phase front, the front velocity rather than the diffusional velocity becomes important. The ratio of the front velocity to the diffusional velocity is

$$\frac{V_n}{(a_l/\lambda_c)} = \frac{\text{St} \lambda_c}{\delta_T} \tag{5.61}$$

Therefore, for small Stefan numbers the front velocity is much smaller than the diffusional velocities. The time scale is chosen for $\mathcal{O}(1)$ motion of the interface as follows. The equation for translation of the interface defined by the curve $f(x, y, t)$ is nondimensionalized to get

$$\frac{\lambda_c}{\tau_l} \frac{d\vec{X}^*}{dt^*} = V_n \vec{n} v_n = \vec{n} v_n \left(\frac{a_l \text{St}}{\delta_T} \right) \tag{5.62}$$

whereby

$$\tau_i = \frac{\lambda_c \delta_T}{a_l \text{St}} \tag{5.63}$$

Note that in this case the parameters δ_T and St need to be input from the macroscopic scale. The key concept here is that we should nondimensionalize the equations according to the scales of interest. Since the solidification front speed for low Stefan numbers is slow compared to the conduction speed, the time scale should be chosen according to the phase change characteristics in order to view the phase change dynamics appropriately.

5.2.5 Pure Conduction

Having decided on the length, velocity, and time scales at the morphological level, we are now in the position to identify the scales of the field variables and to cast the equations at this scale

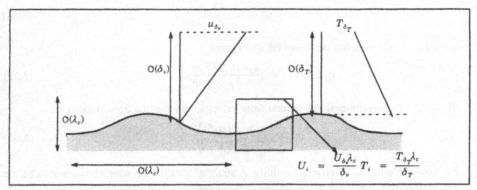

Figure 5.4.   Illustration of velocity and temperature scales at the morphological level.

in the appropriate nondimensional form. At the instability scale, considering the fact that $\lambda_c < \mathcal{O}(\delta_T)$, the temperature scale of concern, as seen in Fig. 5.4, is

$$\Delta_i = \frac{\lambda_c \Delta}{\delta_T} \tag{5.64}$$

Thus, with these scales the equations at the morphological scale take the following forms.

Heat conduction : $$\frac{\lambda_c \, \text{St}}{\delta_T} \Theta_{t^*} = \nabla^2 \Theta \tag{5.65}$$

where $\Theta_i = (T_i - T_m)/\Delta_i$ is the nondimensional temperature. Substituting for λ_c, we get

$$\frac{\pi \, d'}{\sigma' \, \text{St} \, \delta_T} \Theta_{t^*} = \nabla^2 \Theta \tag{5.66}$$

where σ' is a constant appearing in the selection criterion (Udaykumar 1994). For δ_T, an estimate can be obtained from the boundary layer considerations detailed above, or a pure conduction estimate can be obtained.   After substituting for all available scales, the equation for front velocity reads

$$v_n = \frac{\pi^2 \, d'}{2 \, \text{St}^3 \, \delta_T \, \sigma'} \left[\frac{\partial \Theta}{\partial n^*} \right]_{s-l} \tag{5.67}$$

The interface update then takes place according to

$$\frac{d\vec{X^*}}{dt^*} = \left[\frac{\partial \Theta}{\partial n^*} \right] \vec{n} \tag{5.68}$$

which leads to $\mathcal{O}(1)$ interface development in unit time. The Gibbs-Thomson condition is given by

$$\Theta_i = -T_m \left(\frac{\delta_T \sigma'^2 \mathrm{St}^4}{\pi^2 \Delta d} \right) \gamma^*(\phi) \; \varkappa^* \tag{69}$$

where $\Theta_i = (T_i - T_m)/\Delta_i$ is the interfacial temperature, $d = \gamma/L$, and $\varkappa^*$ is the nondimensional curvature of the interface.

5.2.6 Morphological Scales in the Presence of Convection

In determining the scaled equations at the microscopic level, we assume justifiably that $\lambda_c < \mathcal{O}(\delta_v, \delta_T)$. The scales determined are strictly applicable for the diffusion controlled growth only. However, in the absence of preceding work on pattern selection in the presence of convection, we continue to adopt these scales even in the presence of fluid flow. The estimates based on the pure conduction model are expected to hold in the presence of convection at the microscales, since the latter effect is not likely to be strong enough to significantly alter the physics at that scale, at least in the linear stages of growth. In fact, for any Prandtl number the interfacial instabilities are small enough so that the perturbations are embedded within the viscous as well as thermal boundary layers at onset and initial development, until the perturbations assume sizes comparable to the boundary layer thickness. Thus the determination of the scale at incipience is more or less governed by diffusion. In what follows, the ranges of parameters under which each assumption will remain valid will be specified.

In order to obtain the remaining scales in the inner (i.e., morphological) region, we return to the two cases presented previously.

5.2.6.1 Low Prandtl Number Melts

For this case we have

$$\mathrm{Pr} < \mathcal{O}(1) \tag{5.70}$$

Let us assume that

$$\mathcal{O}(\lambda_c) < \mathcal{O}(\delta_v) < \mathcal{O}(\delta_T) \tag{5.71}$$

This usually is the case for most growth environments. The scaling presented here is predicated on this condition. The inequality of the left is necessary for the application of morphological stability theories since these rely on pure conduction transport of heat. Thus, in performing the scaling at the morphological level, it is necessary that the convective influences on the instability be weak. The inequality above holds for all but very small undercoolings. For very low undercoolings, the emerging instability scale λ_c can be quite large (Tirmizi and Gill 1987). It is not clear if convection modifies the linear stability of an unstable interface and thus the pure conduction scales adopted in the following. Substituting for all the known values from Eqs. (5.23, 5.33, 5.57) results in

$$St^2 > O\left[\frac{\pi d' \, Re_d^{1/2}}{\sigma' h}\right] \tag{5.72}$$

This sets a lower limit on the Stefan number. If the Stefan number is very small i.e., the undercooling or front velocity is very small, the length scale of the instability will be large. This implies a slower, thicker primary dendrite. Thus, as the Stefan number is decreased, the length scale of the instability or the dendrite tip radius increases, until in the limit it becomes comparable with a boundary layer thickness. However, there is also an upper limit on the Stefan number. For the scalings to be valid at all, the Peclet number needs to be very small. For small Peclet numbers, the Ivantsov solution reduces to

$$St = \sqrt{\pi \, Pe} \tag{5.73}$$

Since $Pe < O(1)$, this implies that

$$St < O(\sqrt{\pi}) \tag{5.74}$$

Thus the bounds on the Stefan number for this situation can be expressed as

$$O\left(\frac{\pi d \, Re^{1/2}}{\sigma' h}\right) < St < O(\pi) \tag{5.75}$$

The inner scales for the field variable, as seen from Fig. 5.4, are as follows:

$$U_i = \frac{\lambda_c U_g}{\delta_\nu} \tag{5.76}$$

$$\Delta_i = \frac{\lambda_c \Delta}{\delta_T} \tag{5.77}$$

5.2.6.2 High Prandtl Number Melts

For this case,

$$Pr > O(1) \tag{5.78}$$

From the boundary layer considerations, the thickness of the layers are

$$\frac{\delta_\nu}{h} = Re^{-1/2} \tag{5.79}$$

$$\frac{\delta_T}{h} = (\mathrm{Pr}^{-1/3}\,\mathrm{Re}^{-1/2}) \tag{5.80}$$

The scenario pertaining to this case is illustrated in Fig. 5.2 (b), where

$$\mathcal{O}(\lambda_c) < \mathcal{O}(\delta_T) < \mathcal{O}(\delta_v) \tag{5.81}$$

Substituting for the known length scales, simplifying, and considering the lower bound on the Stefan number given by Eq. (5.74), we get

$$\mathcal{O}\left(\frac{\pi\, d\, \mathrm{Re}^{1/2}\, \mathrm{Pr}^{1/2}}{h\sigma'}\right) < \mathrm{St}^2 < \mathcal{O}(\pi) \tag{5.82}$$

The field variable scales are
$$U_i = \frac{\lambda_c U_g}{\delta_v} \tag{5.83}$$

and
$$\Delta_i = \frac{\lambda_c \Delta}{\delta_T} \tag{5.84}$$

The preceding analysis illustrates the wide disparity of the macroscopic, morphological, and microscopic scales that have to be considered in a typical solidification problem. This disparity in scales makes the problem extremely difficult to solve. At the macroscopic scales it is often convenient to cast the equations in a space averaged formulation which smears out the information at the morphological and the microscopic scales. One such formulation is the enthalpy formulation. We will now focus on the enthalpy formulation and the associated computational issues.

5.3 ENTHALPY FORMULATION

In the enthalpy formulation, the energy equation is initially cast in terms of the total enthalpy. By splitting the total enthalpy into the sensible enthalpy and the latent heat, we arrive at a unified formulation to describe energy transport in both phases. The relationship between the sensible heat and the latent heat is modeled in a simplified manner without recourse to either morphological or microscopic scales. In this manner, phase boundaries can be tracked naturally rather than by separate explicit procedures and determined as part of the solution. A spatial average implies that a phase fraction is defined based on the volume fraction occupied by a given phase, say, liquid. Such a two-phase approach is naturally suitable for alloys where phase change generally occurs over a temperature range, according to the local mixture composition, forming a *mushy zone* in which both phases can coexist. A detailed discussion can be found in Crank (1984). For pure materials, which undergo phase change at a given temperature, the mushy zone is an artifact of the modeling procedure. This practice is akin to the shock capturing technique for compressible flows where a finite thickness generally larger than physical reality, is needed to define the shock. However, for both shock and phase change calculations, this artificial smearing is generally quite acceptable and does not necessarily impact the global accuracy of the flow solution. A major requirement for both problems is to devise numerical schemes that are accurate in the bulk of the flow domain and that can resolve the internal boundaries (shock, interface etc.) within one or two computational cells. As will be demonstrated later, in the example of Bridgman growth of CdTe crystals, this requirement can be quite satisfactorily met by the enthalpy formulation.

In the following, we present the salient features of the enthalpy formulation by applying it to a pure conduction problem. Thereafter, its application to a practical problem, including convective effects, will be demonstrated and discussed.

5.3.1 Heat Conduction

In this section we discuss the enthalpy formulation and its application to a heat conduction problem. The governing equation can be cast in the form

$$\frac{\partial(\varrho H)}{\partial t} = \nabla \cdot (k \nabla T) \tag{5.85}$$

where H is the total enthalpy. The equation can be discretized according to the conventional control volume procedure, wherein the total enthalpy H and the temperature T can be interpreted as averaged values within a control volume. Then H can be written as

$$H = C_p T + f \cdot L \tag{5.86}$$

where f is the phase fraction of phase 1 and can be interpreted as the fractional volume of a computational cell occupied by that phase. The phase fraction of phase 1 is zero in the region occupied by phase 2 and unity in the region occupied by phase 1. The phase fraction lies between zero and unity when the control volume is undergoing phase change. L is the latent heat. Substituting equation (5.86) into the governing equation (5.85) we get:

$$\frac{\partial(\varrho C_p T)}{\partial t} = \nabla \cdot (k \nabla T) - L \cdot \frac{\partial(\varrho f)}{\partial t} \tag{5.87}$$

where the latent heat now appears as a source term. This formulation is a single-region formulation, wherein one set of governing equations can describe both phases and hence can more readily handle complex interface shapes, branched and multiple-valued interfaces. To close the physical model, a unique relationship between the fluid fraction f and the temperature T has to be formulated. For a pure substance undergoing isothermal phase change, the total enthalpy, H, is a discontinuous function of the temperature (See Fig. 5.5a)

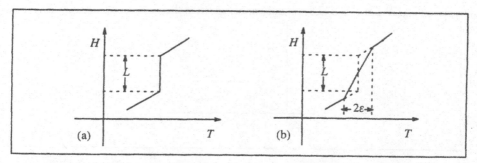

Figure 5.5   (a) Schematic of the enthalpy — temperature relationship for a pure substance; (b) Illustration of the numerical treatment of the discontinuity which is accomplished by smearing the phase change interval over a finite temperature range.

$$H = \begin{cases} C_p \, T \, , & T < T_m \\ C_p \, T + L \, , & T > T_m \end{cases} \tag{5.88}$$

However, from a computational viewpoint discontinuities are difficult to track, and often it is necessary to smear the phase change over a small temperature range to attain numerical stability, as schematically illustrated in Fig. 5.5(b). Thus,

$$H = \begin{cases} C_p \, T \, , & T < T_s \\ C_p \, T_s + \left[C_p + \dfrac{L}{2\varepsilon} \right] (T - T_s) \, , & T_s \leq T \leq T_l \\ C_p \, T + L \, , & T > T_l \end{cases} \tag{5.89}$$

where $T_s = T_m - \varepsilon$, $T_l = T_m + \varepsilon$, and ε is a phase change interval. The discontinuity is thus replaced by a small interval over which phase change occurs. For a pure substance undergoing isothermal phase change, this is a purely numerical artifact and needs to be as low as possible in order to accurately model the physical system. The above relationship is substituted into Eq. (5.5) to obtain:

$$f = \frac{T - T_s}{2 \, \varepsilon} \tag{5.90}$$

which is used to iteratively update the phase fraction from the computed temperature field. This technique has been used successfully by many researchers, including Bennon and Incropera (1987), Lacroix (1989), Lacroix and Voller (1990), Prakash (1990), Prakash and Voller (1989), Shyy et. al. (1992b), Shyy and Chen (1991a), Shyy and Rao (1994a), Voller and Cross (1981),

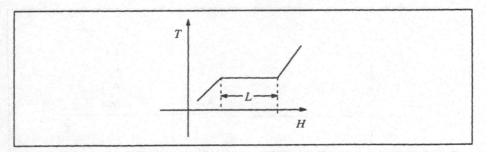

Figure 5.6. Sketch showing the inverted enthalpy — temperature relationship for a pure material.

Voller and Prakash (1987) and henceforth will be referred to as the T-based update method. An alternative formulation is to express the phase fraction as a function of the total enthalpy rather than the temperature, whereby a continuous relationship can be obtained. An implicit procedure is examined here. First, we invert the $H = H(T)$ relationship (5.89) (see Fig. 5.6) so that

$$T = \begin{cases} \dfrac{H}{C_p}, & H < H_s \\[2ex] T_s + \left[\dfrac{2\varepsilon}{L + 2C_p\varepsilon} \right] (H - H_s), & H_s \leq H \leq H_l \\[2ex] \dfrac{H - L}{C_p}, & H > H_l \end{cases} \tag{5.91}$$

where H_s and H_l are the enthalpy values corresponding to the T_s and the T_l temperatures, respectively, $H_s = C_p T_s$ and $H_l = C_p T_l + L$, assuming a constant C_p. Substituting the above relationships into Eq. (5.91), we obtain

$$f = \frac{H - H_s}{H_l - H_s} \tag{5.92}$$

for the iterative update of the fluid fraction. The motivation for the above formulation stems from the fact that the temperature is always a continuous function of the total enthalpy. In particular, it is noted that $\varepsilon = 0$ is allowed in this formulation, thereby accurately modeling phase change of a pure substance. Another motivation is the fact that in the phase change zone, the phase fraction is a rapidly varying function of the temperature, causing it to be sensitive to small errors in temperature. In particular, this can result in an unphysically thick phase change zone (mushy region) due to round-off errors in the temperature calculation. Contrast this with the fact that if the latent heat is large relative to the sensible enthalpy, the phase fraction is a slowly varying function of the total enthalpy resulting in a more stable update procedure. This update procedure will henceforth be referred to as the H-based update method.

5.3.2 Implementation

5.3.2.1 Implementation of the T-Based Method

A straightforward implementation of the T-based method, using the notation given in Chapter 2, would be

$$a_P T_P^k = \sum a_{nb} T_{nb} + b^k - \varrho \cdot L \frac{\Delta x \Delta y}{\Delta t} \left[f_P^{k-1} - f_P^{n-1} \right] \tag{5.93}$$

followed by the iterative update

$$f_P^k = \begin{cases} 0 & , \quad T_P^k < T_s \\ \dfrac{T_P^k - T_s}{2\,\varepsilon} & , \quad T_s \le T_P^k \le T_l \\ 1 & , \quad T_P^k > T_l \end{cases} \tag{5.94}$$

where the superscript k stands for the current iteration value at a given time level n and superscript $n-1$ stands for the previous time level. For Stefan numbers less than unity (large values of latent heat relative to the sensible heat $C_p \Delta T$), this form of the iterative update is prone to numerical instability in the form of oscillations, although for moderate Stefan numbers, underrelaxation of the source term facilitates the solution process. Similar findings have been reported by Prakash et al. (1987), Prakash and Voller (1989), Voller and Prakash (1987). The chief cause of this oscillatory, nonconvergent behavior is the extreme sensitivity of the phase fraction to small changes in temperature, leading to a large negative feedback effect through the source term. This tendency gets worse as the Stefan number is reduced and the time step is refined. A simple hypothetical example will serve to illustrate the deficiencies of the above procedure and also will provide a criterion for devising a stable update procedure. Consider some typical nondimensional parameters: $L = 20$, $T_m = 0.5$, $C_p = 1$. Choose $\varepsilon = 0.1$, then $T_s = 0.4$ and $T_l = 0.6$. Suppose that in the course of the calculation we obtain $T^k = 0.39$, $f^k = 0$ and $T^{k+1} = 0.42$; now calculate the total enthalpy, $H = T^{K+1} + f^k \Delta H = 0.42$; carrying out the update procedure, we get $f^{k+1} = 0.1$. The total enthalpy H jumps to 2.42, an increase of about 500% due to the update procedure alone. This large jump in total enthalpy due to the update procedure is nonphysical; the temperature and fluid fraction fields are no longer consistent with the discrete form of the conservation law. Thus this update procedure will not be able to yield mutually consistent T and f fields and hence is not guaranteed to converge (Shyy and Rao 1994a). A more stable update procedure can now be devised by substituting Eq. (5.94) into the discrete form of the conservation law and moving the resulting coefficients of T_P into the a_P term to give

$$a_P T_P^k = \sum a_{nb} T_{nb}^k + b^k - \varrho \cdot L \frac{\Delta x\, \Delta y}{\Delta t} \left[f_P^{n-1} \right] \quad , \qquad\qquad T_P^{k-1} < T_s$$

$$\left[a_P + \varrho \cdot L \frac{\Delta x\, \Delta y}{\Delta t} \frac{1}{2\varepsilon} \right] T_P^k = \sum a_{nb} T_{nb}^k + \varrho \cdot L \frac{\Delta x\, \Delta y}{\Delta t} \left[\frac{T_l}{2\varepsilon} + f_P^{n-1} \right] \quad , \quad T_s \le T_P^{k-1} \le T_l$$

$$a_P T_P^k = \sum a_{nb} T_{nb}^k + b^k - \varrho \cdot L \frac{\Delta x\, \Delta y}{\Delta t} \left[f_P^{n-1} - 1 \right] \quad , \qquad\qquad T_P^{k-1} > T_l$$

$$(5.95)$$

where the superscript k indicates the current iterative value. It can be observed that with this form of the update procedure, the temperature and the fluid fraction field are computed simultaneously and are therefore mutually consistent at every step of the iterative process. This procedure converges quickly while still retaining accuracy, as will be demonstrated later.

5.3.2.2 Implementation of the H-Based Method

The implementation of the H-based method is relatively straightforward (Shyy and Rao 1994a). Equation (5.93) is applied at every discrete cell, followed by the update procedure, which can be written as follows:

(i) Compute the total enthalpy

$$H_P^k = C_p T_P^k + f_P^{k-1} \cdot L \tag{5.96}$$

(ii) Carry out the update procedure:

$$f_P^k = \begin{cases} 0 & , \quad H_P^k < H_s \\ \dfrac{H_P^k - H_s}{H_l - H_s} & , \quad H_s \le H_P^k \le H_l \\ 1 & , \quad H_P^k > H_l \end{cases} \tag{5.97}$$

5.3.3 Results and Discussion

5.3.3.1 Accuracy Assessment

A one-dimensional conduction driven phase change problem with an exact solution was chosen as the test bed for the assessment of accuracy. Consider an idealized phase change material, with constant thermophysical properties, including density, and with phase change occurring at a constant, specified temperature. This implies the absence of convection, the phase change being driven entirely by the conduction process and release of latent heat at the phase change interface. This class of problems is also known as the Stefan problem. The material is initially at the phase-change temperature, $T_m = 0$. At time, t = 0, the temperature at $x = 0$ is instantaneously raised to $T_h = 1$. The interface is defined by the $T = T_m = 0$ isotherm. For the pure conduction problem in each phase we have

$$\frac{\partial T_i}{\partial t} = \alpha_i \, \nabla^2 T_i \qquad i = \text{liquid, solid} \tag{5.98a}$$

along with the jump condition:

$$k_s \frac{\partial T_s}{\partial n} - k_l \frac{\partial T_l}{\partial n} = \varrho \, L \, v_n \tag{5.98b}$$

where v_n is the interface velocity. For this problem, a closed-form solution exists for the time-dependent location of the phase change front given by $x = 2\lambda \sqrt{\alpha t}$, where λ is the root of

$$\lambda \, \exp\!\left(\lambda^2\right) \, \text{erf}\,(\lambda) = \frac{\text{St}}{\sqrt{\pi}} \tag{4.20}$$

and St is the Stefan number; defined as $\text{St} = C_p \left(T_h - T_m\right)/L$. Note that the heat flux to the interface takes place only through the liquid and the solid remains at a uniform temperature.

Calculations were carried out on uniform grids consisting of 21 and 41 nodes with a time step $\Delta t = 0.01$ and on a grid with 81 points with time steps $\Delta t = 0.1, 0.01,$ and 0.001 for both the H-based and the T-based methods in order to compare the two methods on the basis of accuracy. Comparisons were made with the exact solution to assess the accuracy of the two methods on the basis of the interface location. The interface location was obtained by interpolating for the $f = 0.5$ contour. For the T-based method, $\varepsilon = 10^{-5}$ was adopted. Second-order central differences were used for all spatial derivative terms and first-order Euler backward differencing was employed for marching in time. Figures 5.7(a) and 5.7(b) show that the H-based and T-based methods yield identical solutions even for coarse grids. An error distribution was formed by computing the relative error at every time instant. Figure 5.7(c) shows that the error distribution oscillates about a zero mean. This can be correlated with the fact that a significant part of the latent heat content in a computational cell is released when it starts to undergo phase change. Thus the interface first slows down when a computational cell just starts to change state and then subsequently speeds up. The computed solution coincides with the exact solution when the interface is at the boundary between two control volumes. The amplitude of the error distribution drops with increasing spatial resolution, and its frequency doubles when the grid spacing is halved. A measure of the cumulative error is obtained from the norm of the error distribution, which is the rms value of the error distribution. A log-log plot of the cumulative error versus Δx (Fig. 5.7(d)) shows that the cumulative error varies as $(\Delta x)^{1.5}$.

A calculation conducted with a time step of 0.001 yielded a solution that could not be distinguished from the solution for the time step of 0.01. This indicates that the solution is time-accurate.

5.3.3.2 Performance Assessment

The problem described above was generalized in order to make an extensive comparison of the two update procedures formulated earlier. In general, the solid and liquid phases differ in their thermal diffusivities. Thus the above problem was reformulated such that a steady state solution would exist with the interface positioned within the computational domain so that the effects of heat conduction through the solid would also play an equally important role. So the phase-change temperature was set to $T_m = 0.5$ with all other parameters being the same as the previous study.

   Figures 5.8, 5.9, and 5.10 show the solutions and the relative performance obtained for various combinations of ε, Δt, and Δx. Figures 5.8 and 5.9 show the solutions and performance comparisons on a coarse grid of $\Delta x = 0.04$ for time steps Δt of 0.1 and 0.01 and with ε of 10^{-2}, 10^{-3}, 10^{-4}, and 10^{-6}. As ε decreases, the step-like character of the interface movement becomes obvious. Corresponding to these step-like movements, there are sharp spikes in the iteration count required for convergence at the given time step. It is evident that the T-Based method requires far fewer iterations when ε and Δt are relatively large. It is also noted from Fig. 5.8(d) that the T-Based method performs better than the H-Based method as the interface slows down and approaches the steady state. For a combination of low ε and smaller Δt, the T-Based method does not converge and solutions can be obtained only with the H-Based method as is evident in Fig. 5.8(e) and (f) and also in Fig. 5.9(e) and (f), where $\varepsilon = 10^{-6}$ and a time step $\Delta t = 0.01$ have been employed. Figure 5.10 shows the effect of grid refinement. The spatial resolution was doubled to $\Delta x = 0.02$. In Fig. 5.10(a), for $\varepsilon = 10^{-2}$, the step-like behavior almost vanishes but reappears when ε is decreased to 10^{-3} in (c). The corresponding performance curves in (b) and (d) show the same trends, that is, the T-Based method performs better under more benign conditions, but when ε is reduced the relative advantage reduces and the H-Based method becomes more efficient. The H-Based method is not very sensitive to the value of ε and can work for $\varepsilon = 0$. Also, the T-Based method is more efficient when the interface motion approaches the steady state. In general, it appears that the T-Based method is relatively efficient compared to the H-Based method when the magnitude of the nonlinear term $\dfrac{L\,\Delta x\,\Delta y}{\Delta t}\dfrac{T^n - T^{n-1}}{2\,\varepsilon}$ is sufficiently low. In particular, the performance degrades rapidly with decreasing ε until a point is reached where a solution is no longer feasible with the T-Based method.

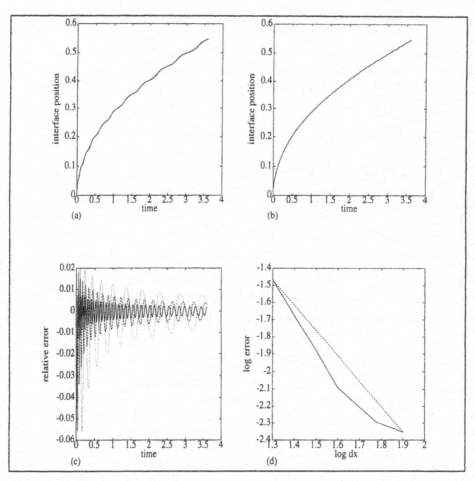

Figure 5.7. Solution features for the pure conduction problem with St = 0.042: (a) Coarse grid solution (21 grid points); (b) Fine grid solution (41 grid points). In each case the solid line designates the solution obtained using the H-based method, the dashed line using the T-based method and the dotted line is the exact solution. The H-based and the T-based method yield identical results. (c) Error at every time instant for various grid sizes. The solid line is the error with $\Delta x = 1/80$, the dashed line is the error with $\Delta x = 1/40$, and the dotted line is the error with $\Delta x = 1/20$. The error oscillates about a zero mean, its amplitude decreases and the frequency doubles when the grid spacing is halved. (d) Cumulative error versus grid spacing for $\Delta t = 0.01$. The solid line is the numerical result, and the dotted line has an approximate slope of 1.5.

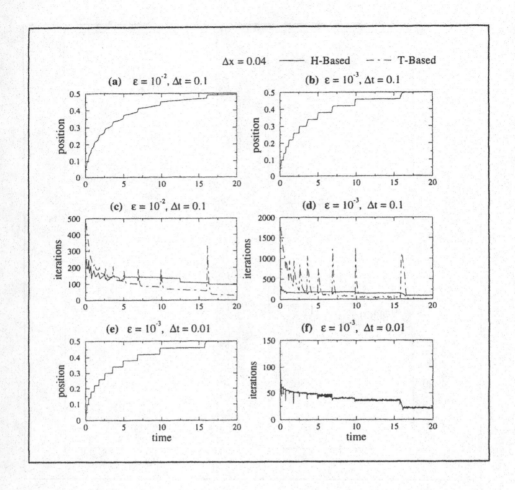

Figure 5.8. *Comparison of the H-Based and the T-Based update methods for the conduction problem using a grid with 26 uniformly distributed points, i.e., $\Delta x = 0.04$. In all cases the solid line represents the H-Based method and the dash-dotted line, the T-Based method. (a) and (b) show the solutions obtained with ε of 10^{-2} and 10^{-3}, respectively. (c) and (d) show the relative performance of the two schemes with ε of 10^{-2} and 10^{-3}, respectively. In (e) and (f), where the time step has been reduced from $\Delta t = 0.1$ to $\Delta t = 0.01$, the T-Based method could not reach a residual of 10^{-4} and has been excluded from consideration.*

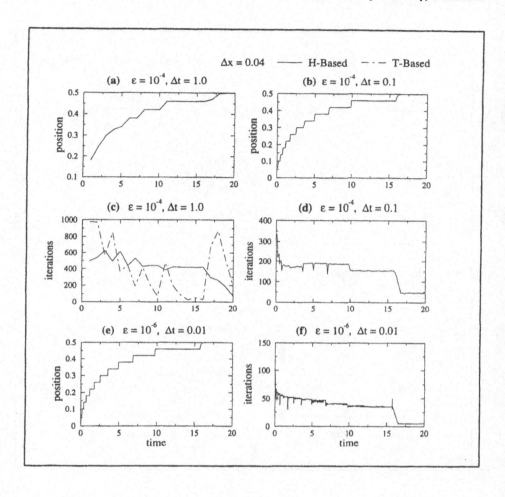

Figure 5.9. *In (a) and (b) the value of ε has been further reduced to 10⁻⁴. It is seen that the T-Based method does not yield solutions when the time step is reduced from $\Delta t = 1.0$ to $\Delta t = 0.1$. When the T-Based method works, it gives better performance than the H-Based method, as shown in (c) and (d). However, from (a) and (b) it is obvious that the solutions are not time step independent. (e) shows the results of the H-Based method for $\varepsilon = 10^{-6}$ and $\Delta t = 0.01$. The T-Based method does not converge to a residual level of 10^{-4} for this combination of lower ε and lower Δt.*

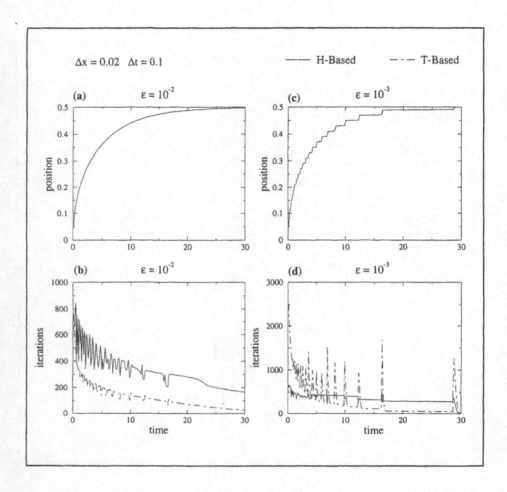

Figure 5.10. This shows the effect of grid refinement. The calculations were carried out on a grid with 51 uniformly distributed points. (a) and (c) show that the step-like behavior of the solutions decreases with increasing spatial resolution. The spikes in the performance curves in (b) and (d) correspond to the step-like movement of the interface in (a) and (c), respectively. The T-Based method performs better in both cases, but its comparative advantage reduces as ε is reduced from 10^{-2} to 10^{-3}. Also, it may be observed that the H-Based method performs better when the temporal gradients are high, i.e., when the interface moves faster due to higher thermal fluxes.

5.3.4   Summary

We have applied the enthalpy formulation to a one-dimensional heat conduction driven phase-change problem to examine alternative procedures for the phase fraction update. The range of applicability and relative performance of the T-based and H-based updates have been examined in the context of solidification problems. It is shown that the T-based update can yield better performance provided the time step is not too small and the grid has sufficient resolution. Recently, Ding and Anghaie (1994) have found that the H-based update yields better performance for the evaporation problem where the boiling point varies with the pressure according to the Clausius-Clayperon equation (Shyy 1994). Thus, the choice of the update procedure depends on the computational parameters, such as time step and grid resolution, and on physical parameters such as the phase change interval and whether the interface temperature varies with the pressure, as in the evaporation/condensation problem. In the next section, we demonstrate the application of the enthalpy formulation to the computation of flow problems using a pressure-based algorithm.

5.4   CONVECTIVE EFFECTS

So far, only conduction-dominated effects have been taken into account, but, as will be demonstrated in the following, convection can substantially modify the flow and heat transfer characteristics. Consequently, the interface shape and translation velocity can be substantially altered by fluid convection. Consideration of fluid convection implies that the fluid fraction field be coupled with the flow solver along with the energy equation. In the energy equation, apart from the fluid convection terms, additional source terms are incorporated. In the momentum equations, the velocities in the solid need to be set to the casting rate of the solid phase. One way to accomplish this is by considering the mushy zone to be a porous medium and incorporating D'arcy law terms in the momentum equations. The D'arcy law terms vary the porosity as a function of the fluid fraction from unity in the liquid to zero in the solid. We now present the governing equations of the enthalpy model written in Cartesian coordinates.

5.4.1   Governing Equations

The governing equations in Cartesian coordinates and in dimensional form for two-dimensional, incompressible flow can be written as:

continuity:
$$\frac{\partial(\varrho u)}{\partial x} + \frac{\partial(\varrho v)}{\partial y} = 0 \qquad (5.99a)$$

x–momentum:

$$\frac{\partial(\varrho u)}{\partial t} + \frac{\partial(\varrho u u)}{\partial x} + \frac{\partial(\varrho u v)}{\partial y} = -\frac{\partial p}{\partial x} + \left[\frac{\partial}{\partial x}\left(\mu \frac{\partial u}{\partial x}\right) + \frac{\partial}{\partial y}\left(\mu \frac{\partial u}{\partial y}\right)\right]$$
$$- A(u - u_{cast}) \tag{5.99b}$$

y–momentum:

$$\frac{\partial(\varrho v)}{\partial t} + \frac{\partial(\varrho u v)}{\partial x} + \frac{\partial(\varrho v v)}{\partial y} = -\frac{\partial p}{\partial y} + \left[\frac{\partial}{\partial x}\left(\mu \frac{\partial v}{\partial x}\right) + \frac{\partial}{\partial y}\left(\mu \frac{\partial v}{\partial y}\right)\right]$$
$$- \left(\varrho - \varrho_{ref}\right) g - A(v - v_{cast}) \tag{5.99c}$$

energy:

$$\frac{\partial(\varrho C_p T)}{\partial t} + \frac{\partial(\varrho C_p u T)}{\partial x} + \frac{\partial(\varrho C_p v T)}{\partial y} = \left[\frac{\partial}{\partial x}\left(k \frac{\partial T}{\partial x}\right) + \frac{\partial}{\partial y}\left(k \frac{\partial T}{\partial y}\right)\right]$$
$$- L \left[\frac{\partial(\varrho f)}{\partial t} + \frac{\partial(\varrho u f)}{\partial x} + \frac{\partial(\varrho v f)}{\partial y}\right] \tag{5.99d}$$

solute transport:

$$\frac{\partial(\varrho \phi)}{\partial t} + \frac{\partial(\varrho u \phi)}{\partial x} + \frac{\partial(\varrho v \phi)}{\partial y} = \left[\frac{\partial}{\partial x}\left(D \frac{\partial \phi}{\partial x}\right) + \frac{\partial}{\partial y}\left(D \frac{\partial \phi}{\partial y}\right)\right] \tag{5.99e}$$

In the above, ϱ is the density, u is the velocity component in the x-direction, v, the velocity component in the y-direction, p is the pressure, μ is the viscosity, k is the conductivity, g is acceleration due to gravity, L is the latent heat of fusion, C_p is the specific heat, f is the liquid fraction, u_{cast} and v_{cast} denote the prescribed velocity at which the solid phase is moving, say, for example, in a crystal growth or a casting process, ϕ is the solute concentration, and D is the mass diffusion coefficient. Here, the gravity vector $\vec{g} = -g\,\hat{j}$ is assumed to be in the negative y direction. For inclined configurations, the appropriate components should appear in both the momentum equations.

5.4.2 Source Terms in the Momentum Equations

When computing the solidification process on a fixed grid, special treatment is needed to enforce the solid velocity to be either zero or equal to the pulling rate of the solid phase in a typical crystal growth or casting operation. The approach taken here is the inclusion of D'arcy-type source terms in the momentum equations. Here the phase change material is considered to be a porous medium, with the porosity changing from zero to unity as the material melts. For pure materials, the porosity changes abruptly as the phase change occurs, but to ensure numerical stability a continuous variation should be imposed. Following Shyy and Chen (1991a), the term A in Eqs. 5.99(b,c) is a function of the liquid fraction as follows:

$$A = C \left(\frac{1 - f^2}{f^3 + q} \right) \qquad (5.100)$$

where the constants C and q are adjusted so that A is at least seven orders of magnitude higher than all the other terms in the momentum equations in the solid region. For a pure material, which melts isothermally, the concept of a porous mushy zone is a numerical artifact, but for alloys, which melt over a temperature range, the porous mushy zone is a physical reality, and convection within the mushy zone definitely influences the structure of the mushy zone (Voller and Prakash 1987). However, it is not straightforward to obtain a relationship between the morphology of the mushy zone and the D'arcy source term, (5.100), and it is certainly a matter for further investigation.

5.4.3 Sources of Convection

Depending on the geometry and the configuration, there can be several driving forces for convection in phase-change systems. Consider typical solidification systems. Unless the solidification process takes place under microgravity conditions, say, on orbiting space platforms, gravity is always present, and thermal inhomogeneities, which also are always present, give rise to density differences. Density gradients also can be caused by compositional inhomogeneities. Depending on the mutual orientation of the gravity vector and the density gradient, buoyancy driven convection can be set up which can significantly alter the temperature and composition distribution within the system and also affect the solid–liquid interface shape and dynamics. If there is a free surface separating the liquid phase and a surrounding ambient gas, a surface tension gradient usually will exist, caused by thermal and composition inhomogeneities, which will exert a shear stress on the fluid phases, leading to Marangoni convection. The different modes of convection can interact to give very complex flow patterns (Shyy and Chen 1993).

Convection also can be caused by density changes during phase change. Usually the different phases have quite different densities. This can be a source of vigorous convection as the fluid moves to conserve mass (Shyy 1994). Many other modes of convection can exist and interact, such as forced convection (due to crystal rotation, etc.), convection due to time-varying electromagnetic fields, etc. In presenting the examples later in this chapter, we will confine ourselves to thermally driven natural convection only.

5.4.4 Computational Procedure

The solution methodology employs a semi-implicit iterative algorithm along with the a finite volume formulation. In the continuity equation (pressure correction equation), first the contravariant velocity components are updated, and then the D'yakonov iteration is used to yield the corresponding Cartesian components in an efficient manner. A staggered grid system is employed to store the dependent variables. Second-order central difference schemes are used to discretize the diffusion terms. For the convection terms, we have a choice of the central difference, first-order upwind or the hybrid scheme. The unsteady terms are discretized according to the implicit first order Euler differencing. The phase-change source terms are discretized in a manner consistent with the rest of the terms in energy equation. Specifically, the $\vec{V} \cdot \nabla f$ term is discretized consistently with the $\vec{V} \cdot \nabla T$ term. Further details of the general solution procedure can be found in Shyy (1994).

5.5 BRIDGMAN GROWTH OF CdTe

The vertical Bridgman configuration is commonly employed to grow single crystals, especially for the electronics industry (Brice 1986, Favier 1990, Vere 1987). The configuration, schematically illustrated in Fig. 5.11, typically consists of a cylindrical ampoule in which the crystal is grown. The ampoule usually is encapsulated by a gas, such as air or argon, or by an inert oxide in liquid form to facilitate translation and/or rotation of the ampoule. During the solidification process, the material is fed into the ampoule and melted and resolidified by varying the temperature field either by translating the ampoule through the furnace or by time-dependent variation of the heater power. The melt/crystal interfacial position and shape will be controlled significantly by the boundary temperature. We now consider the vertical Bridgman growth of Cadmium Telluride (CdTe). CdTe has certain properties which make it important and unique for several applications (Zanio 1978). The combination of its high atomic number, wide bandgap, and good transport properties make it useful for room-temperature gamma-ray and x-ray detectors. The high electrooptic coefficient combined with its low absorption coefficient makes it useful for electrooptic modulators. Low absorption from 1 to 30 μm makes it ideal for lenses, Brewster windows, partial reflectors, and fiber optics. Material made from CdTe is also important in infrared detectors and is used as a substrate for epitaxy, diffusion, or ion implantation. At the present time it has not made as significant an impact as silicon because it is expensive and difficult to grow. For example, its low thermal conductivity ($k \approx 0.01$ W/cm-K) makes the temperature gradient very sensitive to minor variation in input and hence difficult to control, while its low critical resolved shear stress (CRSS≈0.5 x 10⁻⁷ dynes/cm²⁾ makes it very prone to formation of dislocations. The dependence of material properties on temperature distribution and material composition needs to be considered in numerical simulations since it substantially affects the processing conditions and the transport behavior. See Table 5.1 for the material properties of CdTe. Shyy (1994) has identified and discussed several important factors affecting coupled melt/crystal phase change problems and the associated thermofluid transport process including the basic equations that govern this process and the numerical techniques to solve them. The effects of natural convection, capillary force, solute transport, and turbulence on solidification have been examined qualitatively there.

A complete analysis of crystal growth systems encompasses levels of detail ranging from thermal analysis of an entire crystal system to analysis of the dynamics of microscopic morphology in the crystal lattice (Brown 1988). At the system level, the thermal environment affects the entire growth system via heat transfer through the ampoule wall. At the macroscopic transport scale convection and solute segregation in the ampoule influences the formation of the melt/crystal interface shape. Finally, at the morphological level solidification dynamics dictates the formations of cells and dendrites and potential development of crystalline defects. It must be emphasized that these scales interact with one another actively, and a comprehensive model needs to take all these scales into account. The modeling scales of the entire furnace and the melt/solid interface region are not substantially different, and the same transport equations of mass, momentum, and energy apply to both cases. Figure 5.11 shows the schematic of the calculation domain as well as the imposed thermal profile and boundary conditions. The enthalpy formulation facilitates the use of a single set of governing equations to describe both phases, which is sufficient to capture the macroscopic details of the solid-liquid interface shape. The transient Navier-Stokes and associated energy equations in a two-dimensional axisymmetric domain with Boussinesq approximation for treating the buoyancy effect, have been solved with Rayleigh number $Ra = 10^5$ and Prandtl number $Pr = 0.4$. The processing conditions dictate the Stefan number $St = 0.2$.

The following equations, cast in cylindrical coordinates, were solved subject to the boundary conditions shown above, along with the T-based fluid fraction update defined by equations (5.94) and (5.95). The natural convection source terms were linearized according to the Boussinesq approximation:

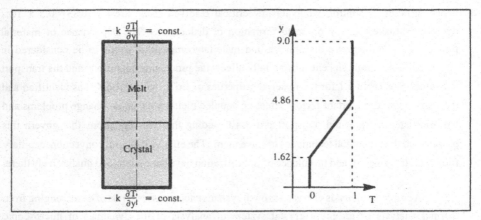

Figure 5.11. Specified thermal conditions on the ampoule wall for the growth of CdTe. Schematic of the growth configuration and boundary conditions.

continuity:

$$\frac{\partial(\varrho u)}{\partial r} + \frac{\partial(\varrho v)}{\partial y} = 0 \tag{5.101a}$$

x–momentum:

$$\frac{\partial(\varrho ru)}{\partial t} + \frac{\partial(\varrho ruu)}{\partial r} + \frac{\partial(\varrho ruv)}{\partial y} = -r\frac{\partial p}{\partial r} + \left[\frac{\partial}{\partial r}\left(\mu r\frac{\partial u}{\partial r}\right) + \frac{\partial}{\partial y}\left(\mu r\frac{\partial u}{\partial y}\right)\right]$$
$$- \frac{2\mu u}{r} - Ar(u - u_{cast}) \tag{5.101b}$$

y–momentum:

$$\frac{\partial(\varrho rv)}{\partial t} + \frac{\partial(\varrho ruv)}{\partial r} + \frac{\partial(\varrho rvv)}{\partial y} = -\frac{\partial(rp)}{\partial y} + \left[\frac{\partial}{\partial r}\left(\mu r\frac{\partial v}{\partial r}\right) + \frac{\partial}{\partial y}\left(\mu r\frac{\partial v}{\partial y}\right)\right]$$
$$- \left(\varrho - \varrho_{ref}\right)rg - Ar(v - v_{cast}) \tag{5.101c}$$

energy:

$$\frac{\partial(r\varrho C_p T)}{\partial t} + \frac{\partial(r\varrho C_p uT)}{\partial r} + \frac{\partial(r\varrho C_p vT)}{\partial y} = \left[\frac{\partial}{\partial r}\left(kr\frac{\partial T}{\partial r}\right) + \frac{\partial}{\partial y}\left(kr\frac{\partial T}{\partial y}\right)\right]$$
$$- L\left[\frac{\partial(\varrho rf)}{\partial t} + \frac{\partial(\varrho ruf)}{\partial r} + \frac{\partial(\varrho rvf)}{\partial y}\right] \tag{5.101d}$$

Table 5.1 Thermophysical properties of CdTe and process parameters

Density	5.67 g/cm³
Thermal conductivity	melt: 0.030 W/cm-K, solid: 0.015 W/cm-K
Heat capacity	melt: 0.187 J/g-K, solid: 0.16 J/g-K
Coefficient of thermal expansion	5×10^{-4} K^{-1}
Dynamic viscosity	0.023 g/cm-s
Latent heat	58 kJ/mol
Melting point	1365 K

Initial calculations were conducted with the material properties held constant and equal to that of the melt phase. Then the thermal conductivity of the melt was elevated to twice that of the solid phase and varied continuously in the mushy zone as

$$k_{mush} = f \, k_l + (1 - f) \, k_s \qquad (5.102)$$

where k_{mush} is the thermal conductivity in the mushy zone, k_l is the conductivity of the liquid phase, and k_s is the thermal conductivity of the solid.

Figure 5.12(i) shows the results for both pure conduction and natural convection calculations with Ra $= 10^5$. The material properties are held constant and equal in both phases for simplicity. With the imposed thermal boundary conditions, the interface is absolutely flat for the pure conduction case and exhibits a small central protrusion for the case with natural convection. The convection is quite weak over most of the domain except in the immediate vicinity of the interface due to the horizontal temperature gradients near the wall. Figure 5.12(ii) shows the corresponding calculations with varying thermal conductivity in the mushy zone. The thermal conductivity of the melt is twice that of the solid phase, which is a good approximation for CdTe. This difference in thermal conductivity changes the isotherm distribution through the solid and the melt. The solid-liquid interface is now concave towards the melt. In this case, convection at Ra $= 10^5$ is not strong enough to have a significant effect on the interface, but the isotherm distribution in the melt is affected by the recirculating flow as observed from (b) and (d) in Fig. 5.12(ii). In all cases, the interface thickness is smaller than one mesh spacing, indicating that the enthalpy formulation can produce solutions with high fidelity. However, if the details of the interface shape play a critical role in determining the solutions, the present approach would face difficulty in incorporating such information. This issue is dealt with in greater detail in subsequent chapters.

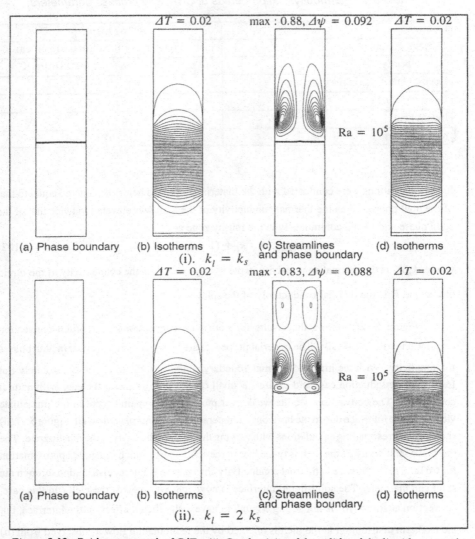

Figure 5.12 Bridgman growth of CdTe. (i) Conductivity of the solid and the liquid are equal. (ii) Conductivity of the liquid is twice that of the solid phase. In each case, (a) and (b) show the mushy zone and isotherms, respectively, for the pure conduction case. (c) shows the streamlines and the mushy zone, and (d) shows the isotherm distribution for a Rayleigh number $Ra = 10^5$. A notable feature is the effect of variable conductivity in the mushy zone, which causes the interface to become concave towards the liquid phase as shown in (ii).

5.6 MULTI-ZONE SIMULATION OF BRIDGMAN GROWTH PROCESS

As previously pointed out, the temperature distribution in the Bridgman system is highly complex and is influenced by several factors such as varying thermophysical properties and multiple heat transfer modes, namely conduction, convection, and radiation. Recent experimental effort in understanding the transport dynamics in the melt can be found in Sears et al. (1992, 1993). These complications traditionally have resulted in simplified analyses carried out under quite restrictive assumptions of simplified geometry, assumed boundary conditions, constant thermo-physical properties and the neglect of important effects such as buoyancy driven convection. The model discussed in the previous section, for the CdTe crystal, addresses only the ampoule portion of the overall growth system. The following model is a more complete one, which accounts for heat transfer in the whole system; including the furnace, the encapsulated fluid between the heater and the ampoule, conjugate heat transfer around and within the ampoule, and phase change dynamics between melt and crystal. Specifically, a two-level approach has been developed to resolve both heat transfer characteristics within the entire furnace and phase change dynamics within the ampoule. At the global furnace level, combined convection/conduction/radiation calculations with realistic geometrical and thermal boundary conditions are made inside the whole system. Refined calculations are then made within the ampoule, with the boundary conditions supplied by the global furnace simulations.

The material considered here is β-NiAl. β-NiAl is an intermetallic that is currently being investigated as a promising high temperature structural material for application in the next generation of aircraft engines and structural components. NiAl is especially attractive because of its low density, high thermal conductivity, high melting temperature, and superior isothermal and cyclic oxidation resistance (Sen and Stefanescu 1991). However, it has two major drawbacks: low toughness at room temperatures and low strength at high temperatures (Darolia 1991). These deficiencies need to be resolved before NiAl can be used in structural applications. Effort has been made to address the various possibilities of improving the properties of this material. In the present context we will discuss the modeling aspects of the solidification dynamics based on recent work by Ouyang and Shyy (1995). In Chapter 2, we discussed various aspects of the transport dynamics in float zone crystal growth process of NiAl. In the following, the Bridgman process is of interest.

Figure 5.13 shows the schematic of a practical design for a vertical Bridgman system currently in use for growing β-NiAl. For the purposes of our calculation, the furnace is composed of two parts:

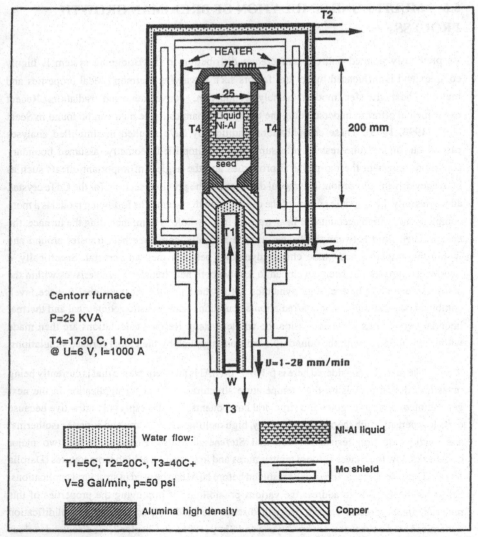

Figure 5.13. Schematic of vertical Bridgman growth furnace for growing β–NiAl.

(i) the enclosure, filled with Argon, containing the heater (T4), and

(ii) the axisymmetric ampoule containing the melt, the growing crystal, and the
     cooler (T1).

The ampoule is made of alumina and rests on a copper support. The enclosure stays stationary during the growth process, and the ampoule is pulled out to control the growth speed of the crystal.

The following phenomena are of key importance and need to be resolved adequately.

- Melt/crystal phase change in the ampoule region
- Conduction among various components of the furnace, the alumina wall, and the copper base
- Convection in the melt and the encapsulated gas argon
- Radiation between the heater T4 and the ampoule wall.

5.6.1 Governing Equations

To formulate this problem with combined heat transfer and phase change in a geometrically complex domain, we have derived and solved a set of unified governing equations that describes conduction, convection, and phase change phenomena over the entire geometry of the system. The dimensionless, axisymmetric, Navier-Stokes, and energy equations, incorporating variable thermal conductivities and the Boussinesq approximation, are:

(i) Continuity equation:

$$\frac{\partial}{\partial r}\left(\frac{\varrho}{\varrho_0}ru\right) + \frac{\partial}{\partial z}\left(\frac{\varrho}{\varrho_0}rw\right) = 0 \tag{5.103}$$

where u and w are the velocity components along the radial and axial direction, respectively

(ii) Momentum equation

r–momentum: $\quad \frac{\partial}{\partial t}\left(\frac{\varrho}{\varrho_0}ru\right) + \frac{\partial}{\partial r}\left(\frac{\varrho}{\varrho_0}ruu\right) + \frac{\partial}{\partial z}\left(\frac{\varrho}{\varrho_0}rwu\right) = -r\frac{\partial P}{\partial r} +$

$$Pr_0\left\{\frac{\partial}{\partial r}\left(\frac{\mu}{\mu_0}r\frac{\partial u}{\partial r}\right) + \frac{\partial}{\partial z}\left(\frac{\mu}{\mu_0}r\frac{\partial u}{\partial z}\right)\right\} - Pr_0\left(\frac{\mu}{\mu_0}\right)\frac{u}{r} + Su \tag{5.104}$$

z–momentum: $\quad \frac{\partial}{\partial t}\left(\frac{\varrho}{\varrho_0}rw\right) + \frac{\partial}{\partial r}\left(\frac{\varrho}{\varrho_0}ruw\right) + \frac{\partial}{\partial z}\left(\frac{\varrho}{\varrho_0}rww\right) = -r\frac{\partial P}{\partial z} +$

$$Pr_0\left\{\frac{\partial}{\partial r}\left(\frac{\mu}{\mu_0}r\frac{\partial w}{\partial r}\right) + \frac{\partial}{\partial z}\left(\frac{\mu}{\mu_0}r\frac{\partial w}{\partial z}\right)\right\} + Ra_0\,Pr_0\left(\frac{\varrho}{\varrho_0}\right)\left(\frac{\beta}{\beta_0}\right)rT + Sw \tag{5.105}$$

(iii) Energy equation:

$$\frac{\partial}{\partial t}\left(\frac{\varrho}{\varrho_0}\frac{Cp}{Cp_0}rT\right) + \frac{\partial}{\partial r}\left(\frac{\varrho}{\varrho_0}\frac{Cp}{Cp_0}ruT\right) + \frac{\partial}{\partial z}\left(\frac{\varrho}{\varrho_0}\frac{Cp}{Cp_0}rwT\right) =$$

$$\frac{\partial}{\partial r}\left(\frac{k}{k_0}r\frac{\partial T}{\partial r}\right) + \frac{\partial}{\partial z}\left(\frac{k}{k_0}r\frac{\partial T}{\partial z}\right) - \qquad (5.106)$$

$$\frac{1}{St_0}\left\{\frac{\partial}{\partial t}\left(\frac{\varrho}{\varrho_0}rf\right) + \frac{\partial}{\partial r}\left(\frac{\varrho}{\varrho_0}ruf\right) + \frac{\partial}{\partial z}\left(\frac{\varrho}{\varrho_0}rwf\right)\right\}$$

where the term with Ra_0 in Eq. (5.3) comes from the Boussinesq approximation for treating the buoyancy effect; Su, Sw in Eq. (5.2) and (5.3) and the term with fluid fraction f in Eq. (5.4) are appropriate source terms to account for the phase change, and ϱ, μ, Cp, k, and β are the density, dynamic viscosity, specific heat, conductivity, and coefficient of thermal expansion respectively. Ra_0, Pr_0, and St_0 are the Rayleigh number, Prandtl number, and Stefan number, respectively. Their definitions are

$$Ra_0 = \frac{\text{Buoyancy Force}}{\text{Viscous Force}} = \frac{\varrho_0 g \beta_0 \Delta T R_0^3}{a_0 \mu_0} \qquad (5.107)$$

$$Pr_0 = \frac{\text{Kinematic Viscosity}}{\text{Heat Diffusivity}} = \frac{\mu_0/\varrho_0}{a_0} \qquad (5.108)$$

$$St_0 = \frac{\text{Sensible Heat}}{\text{Latent Heat}} = \frac{Cp_0 \Delta T}{\Delta H_0} \qquad (5.109)$$

where a_0, ϱ_0, μ_0, Cp_0, k_0, and β_0 are chosen reference material properties for nondimensionalization. g is gravitational acceleration, ΔH_0 is the reference latent heat, R_0 is the reference length, and ΔT is the reference temperature. In this study, we chose the properties of β-NiAl at 300 K as the reference properties; the radius of ampoule inner wall, R_0, as the reference length; and the difference between the highest and the lowest temperature in the furnace, $\Delta T = T4 - T1$, as the reference temperature scale. The estimated dimensionless parameters for our current calculation are

$$Ra_0 = 3.2 \times 10^4 \quad Pr_0 = 4.4 \times 10^{-2} \quad St_0 = 1.7$$

Furthermore, the original parameters of material properties for different material regions, which have been listed in Table 5.2, are converted into the corresponding dimensionless forms, as listed in Table 5.3.

The calculations conducted in this study take into account the variation of material properties with temperature, such as thermal conductivity, at the ampoule walls and at the melt/crystal interface. Abrupt changes in material properties can be expected at the interfaces between two

Table 5.2 Thermodynamic and transport properties				
Material	ϱ (kg/m^3)	μ (kg/m-s)	k (W/m-K)	C_p (J/kg-K)	β (K^{-1})
NiAl	5950	0.005	75	660	1.52 x 10^{-5}
Argon	0.487	5.42 x 10^{-5}	0.0427	520	8.33 x 10^{-4}
Alumina	3975	†	‡	765	–
Copper	8930	†	‡	386	–

† Assigned big values
‡ Varied with temperature (see Fig. 5.14)

Table 5.3 Dimensionless thermodynamic and transport properties				
Material	ϱ/ϱ_0	Pr$_0$ (μ/μ_0)	k/k$_0$	C_p/C_{p0}	Ra$_0$Pr$_0(\beta/\beta_0)$ (ϱ/ϱ_0)
NiAl	1.0	4.4 x 10^{-2}	1.0	1.0	1.4 x 10^3
Argon	8.2 x 10^{-5}	4.77 x 10^{-5}	5.7 x 10^{-4}	0.79	6.32
Alumina	0.67	†	‡	1.16	–
Copper	1.5	†	‡	0.58	–

† Assigned big values
‡ Varied with temperature

media, and the interpolation for the material properties must be handled carefully in order to obtain numerical solutions. The usual approach taken is linear interpolation,

$$k_e = f_e k_P + (1 - f_e)k_E \qquad (5.110)$$

where f_e designates the ratio of distances between points E and e and between points e and P, with E, P, and e representing, respectively, the east side unknown node, the present unknown node, and the east side control surface of the present unknown cell P. However, this does not handle abrupt changes at the interface between two media. Hence, following Patankar (1980), a harmonic interpolation based on a one-dimensional flux conservation is used as follows,

$$k_e = \frac{1}{(1 - f_e)/k_P + f_e/k_E}$$ (5.111)

This treatment has been compared with the flux balance approach (Shyy and Burke 1994) and is found to be equally robust and accurate but simpler to implement. The radiation heat flux between the heat (T4) and the outer wall of the ampoule is handled by the following simplified treatment (Siegel and Howell 1981):

$$q = \frac{\sigma\,(T_1^4 - T_2^4)}{(1/\varepsilon_1) + ((1 - \varepsilon_2)/\varepsilon_2)\,(r_1/r_2)}$$ (5.112)

where $\sigma = 5.67 \times 10^{-8}$ W/(m$^2\cdot$K^4) is the Stefan-Boltzmann constant, T_1, r_1, and ε_1 are temperature, radius, and emissivity for the inner ampoule cylinder and T_2, r_2 and ε_2 are corresponding variables for the heater. The heater is considered to be a blackbody, so $\varepsilon_2 = 1.0$. For the alumina wall of the ampoule, $\varepsilon_1 = 0.8$ and for the copper wall, $\varepsilon_1 = 0.6$. The expression for the boundary condition at the outer ampoule wall is given by

$$- k\frac{\partial T}{\partial r}\bigg|_{ampoule} = - k\frac{\partial T}{\partial r}\bigg|_{Argon} + q$$ (5.113)

where the left hand side represents the heat flux into the ampoule wall and the right hand side represents the heat flux due to convection and conduction from the encapsulating gas and the radiation from the heater.

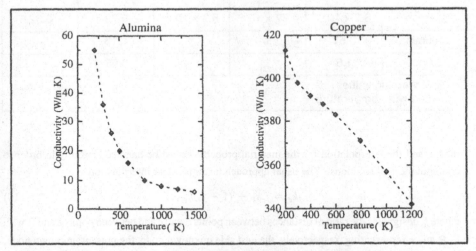

Figure 5.14. Dependence of thermal conductivities of alumina and copper on temperature. (Perry's Chemical Engineer's Handbook 1984)

5.6.2   Two-Level Modeling Strategy

It is observed from Fig. 5.13 that the combination of the furnace enclosure and the ampoule presents a very complicated geometry. To render the computations tractable and simultaneously obtain adequate resolution in the ampoule region, a two-level strategy is employed. Level 1 simulates the entire domain and is referred to as the global furnace model. Level 2 concentrates on the ampoule region and obtains its boundary conditions from the Level 1 or the global simulation. With this two-level strategy, we can obtain useful information at the global level and yet obtain adequate resolution at the melt/crystal interface. Figure 5.15 demonstrates the application of the two-level strategy to the present simulation. We also try to take into account the relative displacement of the ampoule as the growing crystal is pulled out of the furnace.

Therefore, three simulations have been performed at the global level involving three different locations of the ampoule. A multi-zone patched grid method has been employed to generate the grid system needed for global (furnace) level simulations. Sample grid distributions have been shown in Fig. 5.16(a) and (b) for $H = 40$ mm and 80 mm, where H indicates the position of the ampoule within the furnace. The global furnace (Level 1) simulations were conducted for three different ampoule positions in order to estimate the effects of geometric variations as the crystal is pulled out of the furnace. Calculations were conducted for $H = 40$ mm (151×213 grid), 80 mm (151×216 grid) and 120 mm (151×219 grid). In all cases, there were 51×103 points in the ampoule region. Based on this resolution, the ampoule region is computed again with a better resolution; the boundary conditions needed for this Level 2 simulation are extracted from the results obtained from the Level 1 solutions. In the following sections we will first show that consideration of realistic material property variation, as a function of temperature, is essential to obtain a faithful simulation; we will then discuss the solutions obtained for different ampoule locations on both Level 1 and Level 2.

5.6.2.1   The Global Furnace Simulation

Figures 5.17 and 5.18 show the stream function and the isotherm distribution, respectively, for each of the three ampoule locations. The streamline pattern arises as a result of buoyancy driven convection. It is observed that the convection pattern in the encapsulating gas is quite weak and that the convective heat transfer into the ampoule is negligible compared to the radiative effect from the heater. As the ampoule translates downward, the convection within the ampoule gets slightly weaker and the melt/crystal interface becomes convex towards the melt. It is difficult to predict the strength and detailed characteristics of the convective field within the furnace because the geometry is complicated and the boundary conditions vary with the ampoule movement. Nevertheless, it is clear that the isotherm pattern changes quite significantly as the

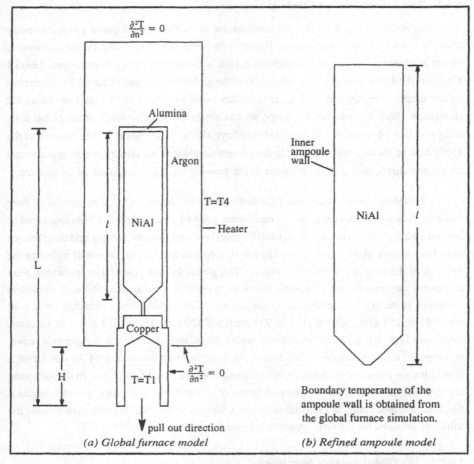

$$\frac{\partial^2 T}{\partial n^2} = 0$$

Alumina

Argon

T=T4

— Heater

NiAl

l

L

Copper

l

T=T1 ← $\frac{\partial^2 T}{\partial n^2} = 0$

H

↓ pull out direction

(a) Global furnace model

Inner
ampoule
wall

NiAl l

Boundary temperature of the
ampoule wall is obtained from
the global furnace simulation.

(b) Refined ampoule model

*Figure 5.15. Layout and boundary conditions for (a) the global furnace model
and (b) the refined ampoule model. H measures the ampoule position as it is
pulled out. The total length of the ampoule between the top of the alumina
crucible and the copper base is L = 185 mm and l = 113.5 mm.*

crystal is drawn down and the temperature gradients at the base of the crystal decrease in
magnitude.

5.6.2.2 The Refined Ampoule Simulation

The refined ampoule (Level 2) simulation was conducted using the ampoule wall temperature
distribution, obtained from Level 1, as the boundary condition. The grid resolution within the

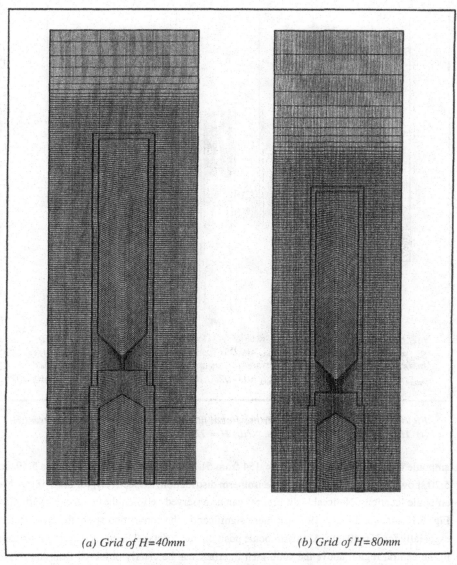

(a) Grid of H=40mm　　　　　(b) Grid of H=80mm

Figure 5.16.　Grid distribution for the global furnace model of (a) H = 40 mm (151 x 213) and (b) H = 80 mm (151 x 216).

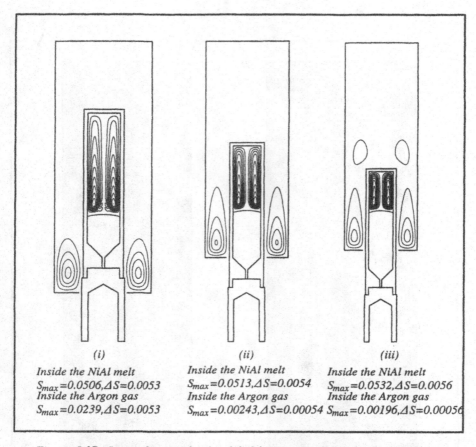

(i)	(ii)	(iii)
Inside the NiAl melt	*Inside the NiAl melt*	*Inside the NiAl melt*
$S_{max}=0.0506, \Delta S=0.0053$	$S_{max}=0.0513, \Delta S=0.0054$	$S_{max}=0.0532, \Delta S=0.0056$
Inside the Argon gas	*Inside the Argon gas*	*Inside the Argon gas*
$S_{max}=0.0239, \Delta S=0.0053$	$S_{max}=0.00243, \Delta S=0.00054$	$S_{max}=0.00196, \Delta S=0.00056$

Figure 5.17. Stream function for the global furnace simulation for three different H.
(i) H = 40 mm, (ii) H = 80 mm, (iii) H = 120 mm.

ampoule has been increased to 101 × 154 from 51 × 103 used at Level 1. Figures 5.19 and 5.20 show the stream function and the isotherm distribution, respectively, for each of the three ampoule locations. Noticeable differences can be observed between the convection patterns in Fig. 5.17(i–iii) and Fig. 5.19(i–iii); more significantly the convection strengths are different, especially for $H = 40$ mm. At this ampoule position, more detailed secondary vortical structure in the central region above the solid-melt interface can be clearly observed from the refined solutions depicted in Fig. 5.19; at the global level, as shown in Fig. 5.17, such characteristics are less pronounced. This assessment shows that a fine grid is needed to capture convection dominated phenomena; thus justifying the two-level approach taken in this study.

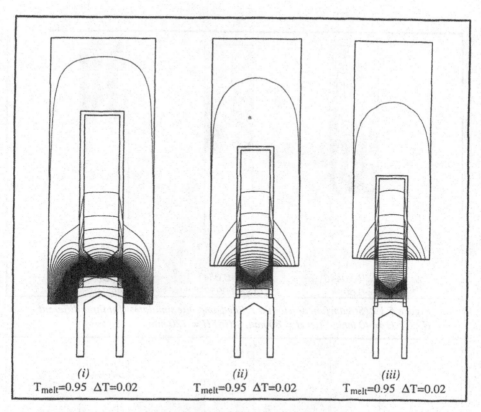

Figure 5.18. Isotherms for the global furnace simulation for three different H.
(i) H = 40 mm, (ii) H = 80 mm, (iii) H = 120 mm.

Figure 5.21 contrasts the melt/crystal interface obtained from the two level simulation for all three ampoule locations. It is observed that the differences between the coarse grid (Level 1) and the fine grid (Level 2) calculations are most significant when the melt volume decreases (as the crystal is pulled out). This observation just reinforces the trends exhibited in the streamfunction plots mentioned previously. The sensitivity to grid refinement increases as the melt volume decreases because of the constraining effect of the top wall. Since the interface positions at the ampoule wall are the same between the two levels, it is significant that the thermal characteristics exhibit differences between the two levels, resulting in different melt/crystal curvatures. This aspect impacts on the detailed distribution of temperature gradients in the interface region, resulting in a different microstructure of the crystal. Correlations between the thermal characteristics in the combined crystal/melt region and the microscopic structure of

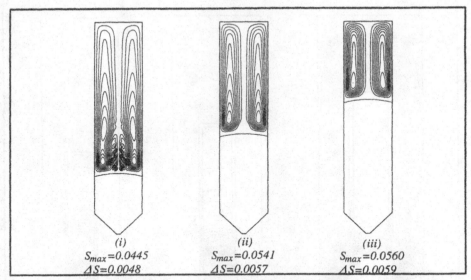

Figure 5.19. Stream function for the refined ampoule simulation for three different H. (i) H = 40 mm, (ii) H = 80 mm, (iii) H = 120 mm

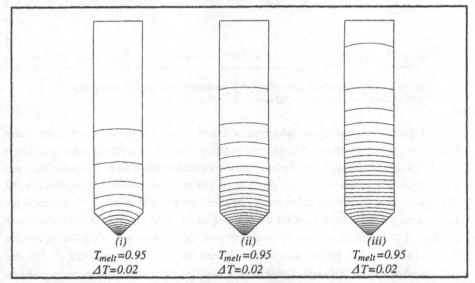

Figure 5.20. Isotherms for the refined ampoule simulation for three different H. (i) H = 40 mm, (ii) H = 80 mm, (iii) H = 120 mm

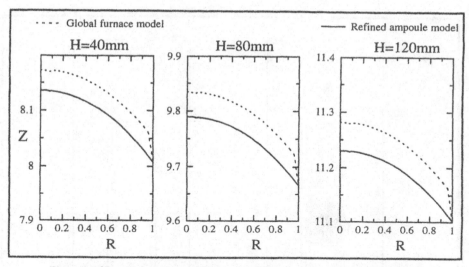

Figure 5.21.   Comparison of the melt/solid interface positions between global furnace and refined ampoule models.

NiAl are not well established at the present time. However, the capability established in this work to predict the solidification dynamics and thermal field will aid the development of such critical information.

To further illustrate the usefulness of the multi–level approach, a fine grid simulation of the $H = 40$ mm case, corresponding to Fig. 5.17(i) and Fig. 5.18(i), was carried on a refined grid of 223×284 nodes (73×154 nodes in the ampoule region). This represents a substantial refinement over the grid system shown in Fig. 5.16(a) with 151×213 nodes (51×103 nodes in the ampoule region). Figures 5.22(i) and (ii) show the streamfunction and isotherms, respectively, obtained on this fine grid system. It can be observed that the fine grid solution captures more details of the convection pattern within the ampoule. Differences can be noted between the solutions presented in Figs. 5.17 and 18 and those presented in Fig. 5.22; more secondary vortical structures are captured on the fine grid. The fine grid solution is closer to the refined ampoule simulation (Figures 5.19 and 5.20) than to the coarse grid solutions from the global furnace simulation. The fine grid results of the global furnace simulation clearly demonstrates the economy of the multi–level approach; instead of resorting to a fine grid distribution at the global level carrying out the refined ampoule simulation yields the desired accuracy economically.

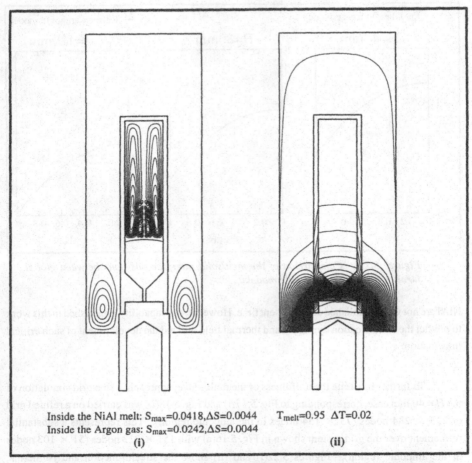

Inside the NiAl melt: $S_{max}=0.0418, \Delta S=0.0044$ $T_{melt}=0.95$ $\Delta T=0.02$
Inside the Argon gas: $S_{max}=0.0242, \Delta S=0.0044$
                                    (i)   **(ii)**

*Figure 5.22.   Solution characteristics of the global furnace simulation for
H=40mm with grid size: 223 × 284. The ampoule region is discretized with 67
× 154 points.*

5.7   FLOAT ZONE GROWTH OF NiAl

In the following we present an additional illustration of the use of the modeling and
computational capabilities discussed in this chapter for practical crystal growth systems. Figure
5.23 shows the schematic of a practical float zone growth system for producing single crystals
of NiAl. The float zone process has been discussed in section 3.2. Here we apply the enthalpy

formulation to a practical growth system as shown in the schematic in Fig. 5.23. This configuration employs induction heating to melt the feed rod of NiAl and maintain the float zone. The induction coil employs a radio frequency alternating current to induce eddy currents within the zone. These eddy currents cause the heat input to the zone via joule heating. If the current in the induction coil is of sufficiently high frequency, then the eddy currents are confined to a thin Hartmann layer adjacent to the free surface. In addition, the induced currents also interact with the external magnetic field, induced by the coil, to exert a normal force on the meniscus and a body force within the melt. In this study, however, we are interested in the convective heat transfer within the melt and its impact on the solid/liquid interface shape. Therefore, the meniscus shape and the heat input will be considered given quantities during the course of the computations.

5.7.1 Calculation Procedure

Issues regarding the formation and stability of the meniscus shapes have been discussed in chapter 2. Equation (2.18) with associated boundary conditions was used to generate the

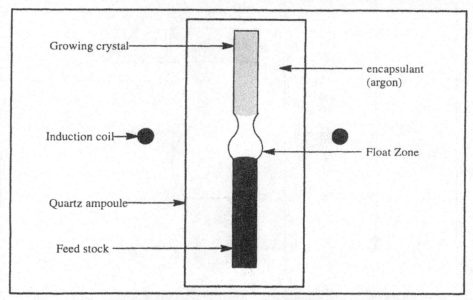

Figure 5.23. Illustration of the float zone process. The float zone is heated by the eddy currents induced by the induction coil and is held up by the surface tension between the melt and the encapsulant. The quality of the crystal is controlled by the thermofluid transport and the shape/velocity of the crystal/melt interface.

meniscus shape. Details of this procedure have been discussed in the example of section 2.3.6 and will not be repeated here. In the current set of calculations, only the heat transfer within the melt and its effect on the melt/crystal and melt/feed interfaces will be of interest. In order to efficiently model the float zone with adequate resolution, it is necessary to consider only half the domain and apply symmetry conditions at the centerline. The governing equations (5.101a—d) and associated boundary conditions are

(i)   Top and bottom boundaries: $\quad \Theta = 0, \ \bar{u} = \bar{v} = 0 \ $ at $ \ Z = 0, H_c$

(ii) Symmetry: $\qquad\qquad \bar{u} \ = \ \dfrac{\partial \bar{v}}{\partial R} \ = \ \dfrac{\partial \Theta}{\partial R} = 0 \ $ at $ \ R = 0$

(iii) Free surface: $\qquad\qquad$ at $ \ R = F(Y)$

(a)   Heat flux: $\qquad\qquad\quad \vec{q} = \nabla\Theta \cdot \hat{n}$

(b)   Marangoni effect: $\qquad\quad \nabla(\vec{V} \cdot \hat{t}) \cdot \hat{n} = $ Ma $ \nabla\Theta \cdot \hat{t}$

$\qquad\qquad\qquad\qquad\qquad\qquad$ (shear stress equals the surface tension gradient)

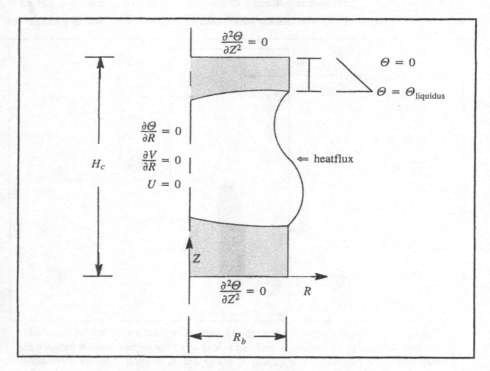

Figure 5.24.   Idealized geometry and boundary conditions.

(c)   Kinematic condition: $\vec{V} \cdot \vec{n} = 0$

(fluid flow is tangential to the free surface)

have been used in the calculations. The geometry and boundary conditions have been illustrated in Fig. 5.24. The governing equations, (5.101a—d) have been solved using the pressure-correction formulation in nonorthogonal body-fitted coordinates. The convection and the diffusion terms in both the momentum and the energy equations have been discretized using second order central differences. The grid points were clustered towards the free surface and for the computations involving Marangoni convection, the grid points also were clustered towards the lower boundary at $Z = 0$. Only steady state calculations have been conducted. Calculations have been carried out on two different grids – a coarse grid of 81 x 51 points and a fine grid of 161 x 101 points – to assess grid independence and solution accuracy. The fine grid system is shown in Fig. 5.25(a). The grid has been clustered towards the free surface to obtain better resolution of the flow and heat transfer characteristics. The results shown in the following are based on the fine grid calculations.

5.7.2   Results and Discussion

It will be shown later in this section that for the given non-dimensional parameter ranges, surface tension gradients generated at the free surface will play a dominant role in melt convection and heat transfer. Based on the thermophysical properties of NiAl and the estimated environmental parameters in the experimental growth configuration of Fig. 5.23, we can arrive at the following non–dimensional parameters.

(i)   Grashof number, Gr = 2000
(ii)   Marangoni number,   100 < Ma < 200   (dominant source of convection)
(iii) Stefan number, St = 0.1, and
(iv)   Prandtl number, Pr = 0.1

The definitions of the parameters listed above are based on those discussed in section 2.3. The heat flux is assumed to be input normal to the free surface and is estimated based on methods outlined in section 2.3.

5.7.2.1   Heat Conduction

A calculation was carried out with the heat transfer taking place through conduction effects only. The purpose was to carry out an *a posteriori* verification of the temperature scales used to define the Grashof and the Marangoni numbers in subsequent calculations. It is noted that due to the curved geometry of the free surface, the location of the maximum temperature shifts towards the convex portion of the free surface, region (B), as is evident from Fig 5.25(b). The

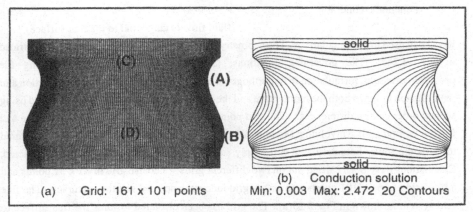

Figure 5.25   (a) Illustrates the grid used. (A) denotes the region adjacent to the
concave side of the free surface; (B) denotes region adjacent to the convex side of
the free surface; (C) and (D) denote the regions close to the intersection of the
solid/liquid interfaces and the centerline as shown.
(b) shows the conduction solution. The solid/liquid interfaces are mildly convex
towards the melt in regions (C) and (D) at the centerline and concave close to the
free surface at regions (A) and (B).

nondimensional value of Θ_{max} is 2.47. The solid/liquid interfaces show moderate curvature at
the axis of symmetry (regions C and D) as is evident from Fig 5.25(b).

5.7.2.2   Thermocapillary Convection

Calculations were conducted for Gr = 2000 and two Marangoni numbers, 100 and 200. Figures
5.26(a) and (b) show the streamfunction and isotherms respectively for the case of Gr = 2000
and Ma = 100. For these values of the parameters, the surface tension gradient, in the form of
Marangoni convection, plays a more dominant role than buoyancy, in the form of natural
convection, in determining the heat transfer characteristics. Figures 5.27(a) and (b) show the
corresponding quantities for Gr = 2000 and Ma = 200. It is evident that as the Marangoni number
increases, the solid/liquid interfaces show increased convexity towards the melt, especially near
the axis of symmetry, i.e., in regions (C) and (D). A clockwise convection cell is formed in
region (B), adjacent to the convex side of the free surface and an anticlockwise cell in region
(A) adjacent to the concave side of the free surface. The location of the maximum temperature
is pulled downwards towards region (B) by the stronger recirculation of the clockwise
convection cell, which in turn is caused by the stronger surface temperature gradients there. As
the convection strength increases with the Marangoni number, the magnitude of the maximum
temperature decreases due to the increased heat transfer effectiveness. Increasing the strength

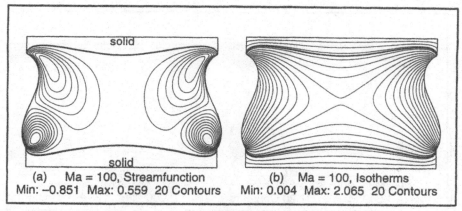

Figure 5.26   Ma = 100. (a) Streamfunction: The bold lines indicate the solid/liquid interfaces. The increase in convexity of the interfaces in regions (C) and (D) is evident. There is a clockwise convection cell in region (B) adjacent to the convex side of the free surface and a counterclockwise cell adjacent to the concave side of the interface in region (A). This is because the surface tension decreases with temperature. From (b) the maximum temperature occurs close to the point of inflexion of the free surface. The convex region of the free surface leads to a stronger convection cell.

of Marangoni convection causes the heat transport to be tangential to the free surface. As a consequence of the decreased heat transfer in the normal direction, the centerline experiences a lower temperature. The increasing temperature inhomogeneity due to this type of convection pattern results in increased curvature of the solid/liquid interfaces and decreasing volume of the melt. The trend is clearly demonstrated by the centerline temperatures shown in Fig. 5.28(a), which shows that the centerline temperatures decrease with increasing Marangoni number, and from Fig. 5.28(b), which shows increasing curvature of the solid/liquid interfaces as a result. The overall effect of Marangoni convection is to increase the temperature inhomogeneity of the melt, especially in the radial direction, causing the solid/liquid interfaces to experience increased curvature towards the melt.

The above series of calculations demonstrate the application of the enthalpy method and its effectiveness in treating problems with complex geometry and involving the interaction of the solidification interfaces with a deformable free surface. It is shown that surface tension gradients generate strong convection in the vicinity of the free surface which significantly affects the heat transfer characteristics and interact with the phase change dynamics. The resulting

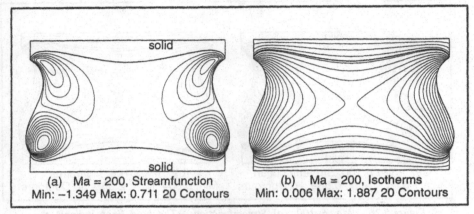

(a)   Ma = 200, Streamfunction
Min: −1.349 Max: 0.711 20 Contours

(b)   Ma = 200, Isotherms
Min: 0.006 Max: 1.887 20 Contours

Figure 5.27   Ma = 200. The convex region of the free surface leads to a stronger convection cell in region (B). Therefore the clockwise cell is much stronger than the counterclockwise cell in region (A). The location of maximum temperature is pulled downwards towards (B) by the clockwise convection cell. The magnitude of maximum temperature has decreased significantly due to increased heat transport by the convection cells. Decreased heat transfer to the center caused increased convexity of the solid/liquid interfaces near the centerline and increased distortion near the free surface due to the tangential heat transport at the free surface.

increases in the curvature of the crystal/melt interface will have a significant impact on crystal quality.

It should be pointed out that both analytical and computational investigations of float-zone crystal growth and related transport processes have been attempted by many researchers. Investigations of pure Marangoni convection, combined Marangoni and natural (Rayleigh) convection, and the interaction of Marangoni-Rayleigh convection with morphological instability can be found in Chen et al. (1994), Lan and Kou (1991a, b), Nadarajah and Narayanan (1986, 1987, 1989, 1990), Narayanan et al. (1992), Quon et al. (1993), Tao et al. (1995), Wagner et al. (1994), and Young and Chait (1990). Related experimental information of thermocapillary effects and Marangoni convection can be found in Chun (1984), Kamotani et al. (1994), Ostrach et al. (1993), and Preisser et al. (1983). In particular, the recent book by Koschmieder (1993) gives a comprehensive account of both analytical and experimental aspects of Marangoni-Rayleigh convection.

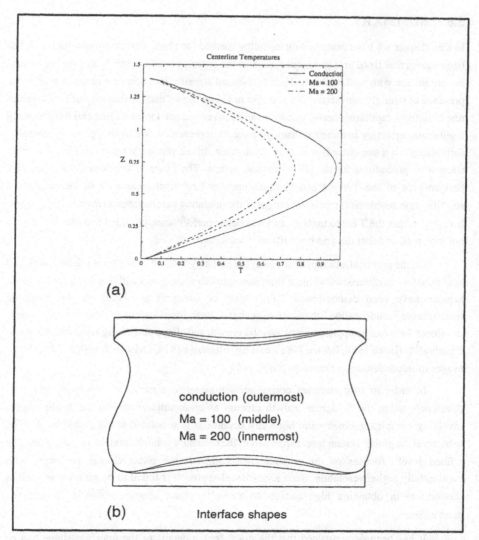

Figure 5.28. Centerline temperatures are reduced in the melt as the Marangoni number increases as is evident from (a). The solid-liquid interface is drawn inward towards the melt at the centerline but pushed towards the solid near the free surface as the Marangoni number increases, as shown in (b). Marangoni convection causes the heat flux to move tangentially to the interface and decreases the heat transported to the interior. This causes the interface to distort and become convex towards the melt, as the Marangoni effect becomes stronger.

5.8   SUMMARY

In this chapter we have described the enthalpy method for phase change computations. It has been shown that fixed grid techniques, such as the enthalpy method, can be conveniently used in conjunction with well established pressure-based algorithms to compute phase change in the presence of strongly convective flows. It also has been shown that the disparity of length scales resulting from capillarity, heat conduction, and convection need to be treated carefully. Scaling arguments pertaining to these mechanisms have been presented. We have applied the enthalpy formulation to a one-dimensional heat conduction driven phase change problem to develop alternative procedures for the phase fraction update. The range of applicability and relative performance of the T-based and H-based updates have been examined in the context of solidification problems for different ranges of the operating parameters and material properties. It is shown that the T-based update can yield better performance provided that the time step is not too small and that the grid has sufficient resolution.

On the practical side, applications of the enthalpy method to compute solidification and heat transfer characteristics within a float zone growth system and within a vertical Bridgman furnace have been demonstrated. The effects of transport properties on the predicted macroscopic solidification characteristics have been demonstrated. The methodologies developed here can be applied to other crystal growth and alloys processing techniques, such as Czochralski (Brice 1986, Brown 1988), casting (Flemings 1974, Shyy et al. 1992b, 1993c), and evaporation/condensation problems (Shyy 1994).

In order to treat complex crystal growth systems, a multilevel approach has been developed, using the Bridgman growth process as an example, wherein the entire system involving complex geometry and boundary conditions is simulated at the global level. The solution at the global system level forms part of the boundary conditions at the refined level. The refined level   focuses on the ampoule and computes the phase change processes with substantially higher resolution. Such a multilevel approach is found to be accurate as well as economical in obtaining high quality solutions to phase change problems in complex geometries.

It has been demonstrated that the macroscopic details of the interface shape can be captured by the enthalpy formulation implicitly, maintaining good global accuracy. However, because of the smearing of information involved in the spatial averaging, the curvature of the interface cannot be accurately obtained. Furthermore, because of the averaging process involved in representing the phase-change source terms, the interface trajectory often exhibits a staircase phenomenon. This implies that certain physical phenomena, such as the Gibbs-Thomson effect, cannot be captured in a straightforward manner. Approximate techniques such as the Volume of Fluid (VOF) methods can be conveniently incorporated into this algorithm. Recent developments aimed at modeling surface tension effects at the continuum level (Brackbill et al. 1992) have made such a combination more accurate and efficient. Nevertheless, curvature dependent internal boundary conditions are still treated approximately, and the conservation laws are not enforced at precisely defined interfacial locations.

In the next chapter we describe a procedure which uses a Lagrangian technique to explicitly track the interface, thus retaining precise interface definition while utilizing the power of the fixed grid, pressure-based algorithms to obtain solutions to the thermal and flow fields. Such a combined Eulerian-Lagrangian computational technique can be applied to a variety of physical problems, ranging from microscopic phenomena controlled by capillarity and conduction processes to macroscopic fluid flows with high Reynolds number and complex geometry.

FIXED GRID TECHNIQUES: ELAFINT—EULERIAN-LAGRANGIAN ALGORITHM FOR INTERFACE TRACKING

6.1 INTRODUCTION

In Chapter 4, we followed the evolution of an unstable interface and illustrated the effects of surface tension on interfacial characteristics. There, we employed marker particles to define the interface. Joining the particles with parabolic segments enabled us to obtain accurate values of the curvature and normals at each point on the interface. A body-fitted grid (Shyy 1994, Thompson et al. 1985) and generalized curvilinear coordinates were employed to map the irregular phase boundary onto a regular computational domain. The boundary condition on the interface was, in this way, applied at the exact interfacial location. However, it was evident that when the interface deformation became severe, an adaptive grid method would experience serious difficulties. In the first place, generating a suitable grid would be inconvenient. Mesh skewness and grid stretching effects can compromise the solutions. Distribution of grids in preferred regions could also be complicated. In the event of topological changes of the interface, readjustment of grids would involve intense logical manipulations, including various interpolations, leading to inaccuracies. On the other hand, as demonstrated in Chapter 5, a purely Eulerian strategy, with a fixed grid can aid in following the motion of the front. In this method, the interface information is not available explicitly, but is extracted, as required, from the information regarding a field variable, such as fluid fraction. Most Eulerian techniques, such as the level-set algorithm (Osher and Sethian 1988, Sethian and Strain 1992) or the phase-field

(Kobayashi 1993, Wheeler et al. 1993) approach described in Chapter 1, and the enthalpy formulation used in Chapter 5, rely on deducing the interface shape based on such information regarding fractional content of some advected quantity in a grid cell.

Hitherto, the computations of the full phase-change problem including the Navier-Stokes equations have been restricted to the macroscopic scales. The most popular methods employed for the computations are purely Eulerian in nature, and employ variants of the enthalpy method described in Chapter 5 (also see Lacroix and Voller, 1990, Shyy and Rao 1994a, Voller and Cross 1981, Smith 1981). The interfaces are then deduced from the contours of a transported scalar, such as fluid fraction. The interface is imparted a thickness in order to overcome the classical difficulty in dealing with sharp gradients/discontinuities in an Eulerian framework. Thus a porosity is assigned in the vicinity of the solid-liquid interface and a D'arcy type term is included to smooth the drastic jump in viscosity across the interface, as discussed in Chapter 5. Thus, unless a very fine grid is employed in the vicinity of the interface, it is not possible to track a sharp phase front by these techniques. At the level of the macroscopic interfaces, the purely Eulerian methods are indeed useful for the following reasons:

1.   The macroscopic interface is unlikely to be highly distorted. Thus the interfacial boundary conditions, which are functions of the interfacial curvature and slope such as the phase-change temperature and concentration jump at the phase front, are unlikely to significantly impact the global thermal and velocity fields. In other words, the capillarity effect given by the Gibbs-Thomson condition, Eq. (5.1), is unlikely to play an important role for small to moderate interface curvatures.

2.   For most materials containing impurities, there exists a mushy zone, which represents a region where the solid and liquid coexist. The information regarding the microstructure is averaged over this region. If one is not concerned with the detail of the microstructure, the interface can be smeared over this mushy zone. This type of treatment lends itself to the purely Eulerian treatment where the interface must have a finite thickness.

3.   The Eulerian methods also easily handle topological changes because the interface is captured rather than tracked. However, the accuracy of the computations in such events cannot be assessed easily.

When the focus is on interfacial behavior at the microscopic levels, the issues that assume importance are capillarity, interfacial distortion, and interfacial motion. The interface then has to be tracked as a discontinuity, which limits the applicability of a purely Eulerian scheme. It would therefore seem worthwhile to explore the possibility of combining the strengths of the Lagrangian and Eulerian strategies, and of devising a method which operates along the

following lines:
1. Employ a fixed grid, thus surmounting grid motion and rearrangement problems.
2. Retain explicit, piecewise functional information of interface shape to the desired order of accuracy.
3. Handle the topological changes of the interface.
4. Devise a robust field solver which honors the governing conservation laws and also handles the interfacial motion and boundary conditions.

With these broad guidelines, the simulation of highly deformed interfaces can be facilitated by making two major modifications to the solution strategy. Since the immediate focus of our effort is to be able to resolve interfaces of possibly highly branched nature, our previous strategy of using body-fitted coordinates to conform exactly to the interface will need to be abandoned. However, the broad outline of the interface tracking procedure is maintained and extended in this chapter. The method to be developed herein is designed to track the interface over an underlying fixed Cartesian grid, such that the interface does not necessarily correspond to a grid line. Thereby, a boundary-conforming grid is rendered unnecessary. However, the fact that the interface lies within a grid cell will necessitate other measures in the solution of the field equations, a key issue to which we will denote considerable attention in this chapter.

ELAFINT (Eulerian–Lagrangian Algorithm For INterface Tracking) is an approach designed with such guidelines in mind. On the Lagrangian side, a methodology is devised to track highly distorted interfaces and their interactions based on the notion of massless markers. This information regarding the interface is then combined with a flow solver which is cast in a stationary grid system. The mixed Eulerian-Lagrangian, time-implicit formulations of the method are maintained and the fluid flow equations are discretized. In section 6.3.2.1 we describe the application of a finite volume formulation for a general transport variable ϕ and the discretization scheme. In sections 6.3.2.3 to 6.3.2.13, we clarify the treatment of the individual terms for each of the flow variables and explain the various issues involved in relation to conservation and consistency. We cast the equations in the computational notation of the pressure-based methodology. We then present some test cases and comparisons to validate the solution procedure. In the course of validating the present algorithm, solutions obtained by other techniques will be employed as reference; such an exercise will be presented in the next chapter.

6.2 INTERFACE TRACKING PROCEDURE

As far as the interface tracking is concerned, the conditions engendered by the typical interfacial instability, which we seek to address via our numerical procedure, may be listed as follows:
1) The interface is likely to be highly branched. There also exists the possibility of fragmentation

or merger of interfaces. The generation of a body-fitted grid to conform to the interface shape would then be impractical.

2) The equations governing the instability phenomena depend critically on such features of the curve defining the interface as the local normal and curvature. These quantities appear in the form of derivatives of the defining curve $F(X,Y,t) = 0$ for the interface. The first and second derivatives F_x, F_y, F_{xx}, F_{yy} need to be obtained very accurately. The calculation of the derivatives is prone to numerical inaccuracy, since these are obtained by the division of small numbers by other small numbers. Their accuracy is nevertheless of utmost importance for a successful numerical calculation.

The above primary restrictions influence the choice of numerical scheme for interface tracking. Additionally, a robust numerical procedure is desired, which does not, in its implementation, further complicate an already formidable problem. We require a simple interface update procedure that is relatively inexpensive. In the marker-based method:

(a) The number of data points involved in interface tracking equals the number of marker particles. No additional equations are to be solved for each cell of the domain as in the VOF (Hirt and Nichols 1981, Liang 1991) or the level-set method (Sethian and Strain, 1992).

(b) There is no ambiguity regarding interface location as in the VOF method. Crisp, smooth interfaces can be obtained and slopes and curvatures calculated accurately.

(c) No elaborate logical operations are involved in obtaining an updated interface shape.

Hence, we have chosen to develop a grid-supported marker-particle scheme to enable us to simulate the dynamics of interface development. In what follows, to facilitate the presentation of the methodology developed, the domains defined by the interface are termed solid and liquid, with the particular instability of a solidifying front as a working example. The applicability of this tracking procedure is, of course, not limited to any particular physical phenomenon. More detailed discussion can be found in Udaykumar and Shyy (1994a) and Shyy (1994).

6.2.1 Basic Methodology

There are three essential elements to our interfacial tracking procedure, namely, cells, interface particles, and interfacial curves connecting the marker particles. The interfacial particles define the terminal points of the interfacial curves. Daly (1967,1969) employed a cubic spline fit to define the interfacial curves. Due to many advantages, circular arc elements are used in this work to represent interfacial segments. Thus, one expresses the interfacial curves as $(X_i - A_i)^2 + (Y_i - B_i)^2 = R_i^2$ where R_i is the radius of the osculating circle at point i and A_i and B_i are the coordinates of the center of the osculating circle. The calculation using circles enables evaluation of normal and curvature in a very simple fashion. The determination of the direction

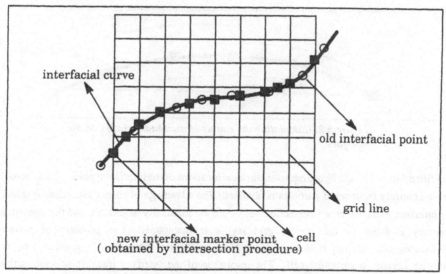

Figure 6.1. Definition of new interfacial points using the intersection procedure.

○   *initial markers used to advance the front*
■   *new markers resulting from redistribution via intersection procedure*

assumed by the normal, which is required to point from the solid into the liquid, is also facilitated and will be discussed shortly. Furthermore, the use of circular arcs is especially attractive when the estimation of local curvature is needed. With circular arcs representing the interface shape, constant curvatures will be maintained between any two consecutive marker points, which is consistent with most of the numerical schemes for the solution of the governing field equations, such as the heat transport equation.

The use of cells, in line with Miyata (1986), offers significant advantages. The procedure employed is illustrated in Fig. 6.1. Let the open circles in Fig. 6.1 represent an initial distribution of interfacial points or the interfacial point locations obtained after advancing the interface at any given time step. The points are found lying within the cells. These interfacial points are connected using circular arcs as follows:

For three given points, such as shown in Fig. 6.2, we obtain A_i, B_i, R_i by elimination among the relations $F(X_i, Y_i, t) = 0$, $F(X_{i+1}, Y_{i+1}, t) = 0$, and $F(X_{i-1}, Y_{i-1}, t) = 0$. The intersection of each segment of the interface with the grid lines is then obtained in a straightforward manner.

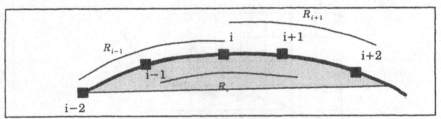

Figure 6.2. Calculating the radius of osculating circles on the interface.

The filled squares on the figure represent the new locations of the interfacial points. Thus, at each step an entirely fresh set of particles is obtained. The advantage of such a procedure is that as the interface expands, new intersection points are automatically generated, and the numerical accuracy dictated by the current grid layout is maintained. The problem of particle rarefaction/accumulation is thus avoided. New points are automatically generated as the interface invades a particular cell. The resolution of the interface tracking scheme is thus determined by the grid spacing. However, this is not necessarily a limitation, since the interfacial point motion is governed by the gradients of the thermal field and the maximum expected resolution is in effect controlled by the resolution of thermal gradients, i.e., the grid spacing. Thus the improved resolution obtained by using several interfacial points per cell is merely artificial. If further resolution is necessary in any region of the interface, adaptive regridding can be performed to improve both the interface and field variable resolution. The other advantage with locating interfacial particles on cell faces is that this facilitates the computation of interfacial velocity and the assignment of boundary conditions and control volumes corresponding to cell control points. This information is necessary anyway and no superfluous operations are introduced by the simple intersection procedure. Once the new interfacial points are determined and they are again connected via interfacial curve segments, directions of the normals can be determined.

Consider the shape of the interface as shown in Fig. 6.3. The osculating circles at different points on the interface are shown. If one were to obtain the normal to the circle at a point, one would compute

$$\vec{n} = \frac{(X_i - A_i)\vec{i} + (Y_i - B_i)\vec{j}}{R_i} \tag{6.1}$$

The directions of the normal to the osculating circles at the sample points are shown along with the required normal direction, which is designated by convention to point from the solid to the liquid. Also, at points such as 1 and 5, the curvatures would erroneously be indicated as being positive (since R_i is always positive), while the expected values are negative, as viewed from the solid. Therefore, a decision as to the sign of the normal and curvature is required to be made. The criterion that has been found to work well in determining whether to invert the sign of the normal is outlined here:

(1)   Nomenclature: The interfacial points are first assigned numbers in sequence such that the solid lies to the right as one advances along the interface, as shown in Fig. 6.4. When renumbering interfaces after mergers, this convention is adhered to.

(2)   Decision regarding inversion: Four basic types are identified and are displayed in Fig. 6.5(a) –(d). In cases A and D, the signs of the normal and curvature are inverted. In cases B and C, the signs are retained. This is in line with our interfacial point numbering convention,

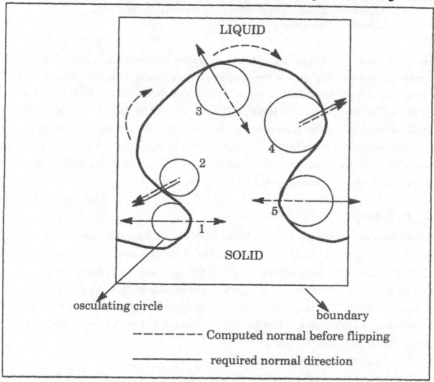

Figure 6.3. Illustration of computed and required normal directions.

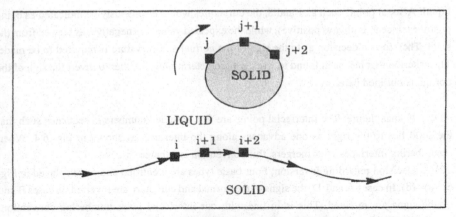

Figure 6.4. Nomenclature of interfacial points. Arrows show
direction of numbering.

whereby the solid lies always to the right as we traverse the interface. In case A, for instance, the center of the osculating circle at the point i lies in the liquid. This is so since $X_i > A_i$. For this case, $Y_{i+1} > Y_{i-1}$. The normal, therefore, points from the liquid into the solid and needs to be reversed. The criterion for case B is $(Y_{i+i} > Y_{i-1}, X_i < A_i)$. In such a case, the center of the circle lies in the solid, and no reversal is necessary. Cases C and D may be argued similarly. It is important to adopt these conventions, in order to remove any ambiguity in defining the phases adjoining the interface. These procedures are hereafter followed in all the tracking procedures used in the case studies presented in the following sections.

6.2.2 Procedures for Mergers/Breakups

Another advantage of the use of cells in the interface tracking procedure is related to the treatment of mergers and breakups. The treatment of mergers has received little attention in connection with marker-based methods, since it has been thought to involve intense logical manipulation. This is indeed the case in the absence of an underlying grid, such as in body-fitted coordinates. These difficulties are mitigated considerably when one associates with each point of the interface, a cell number, as detailed below. Let us define the following data arrays:

For each interfacial marker particle we record

Interfacial marker particle number: i

Index of cell to which the point belongs (x and y index of cell): *cellx* (i), *celly* (i)

After having found the cell number in which a particular interface point lies, that cell is declared as an interfacial cell. Thus, cells in the bulk are identified by setting *cell_flag(i,j)* =0,

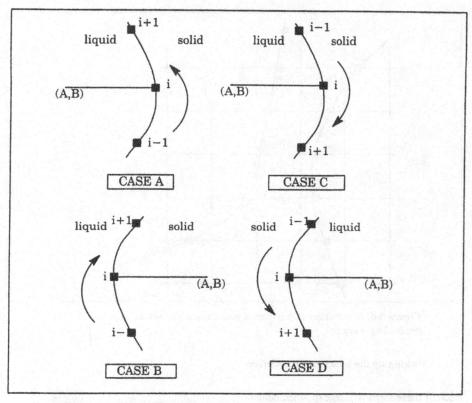

Figure 6.5. Illustration of conditions for switching of normals.

while cells containing interface segments are flagged *cell_flag(i,j)=1*. Conversely, with each cell, the associated point numbers are obtained. This is achieved by sorting through the interfacial cells and finding the identity of the interfacial markers that lie in the cell. Thus,

cell number (x, y): *cellx* (i) , *celly*(i)

interfacial point number in that cell: *pointnum(cellx*(i), *celly*(i), *n*), where *n* is the number of markers in that cell.

The use of *pointnum*, which contains the integer value indicating the index of the marker particle in that cell, enables us to reduce the number of sorting operations when the interfaces interact. There are three steps in the merger procedure:

(1)   Detecting merger locations,

(2)   Setting up for executing mergers,

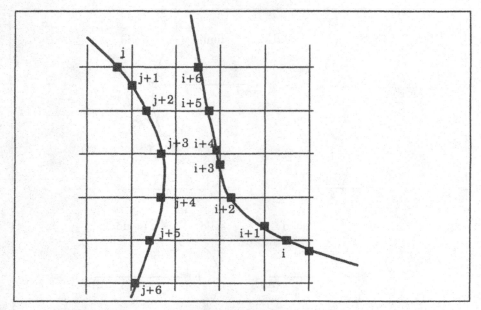

Figure 6.6. Illustration of interfacial point arrangement during impending merger.

(3)   Picking up the new merged interface.

Each of these steps will be explained below.

Step 1. Detecting mergers: Equipped with the information outlined above, the merger locations are detected. Consider elements of the interface as shown in Fig. 6.6. In Fig. 6.6, the markers (j+1 to j+4) and (i+6 to i+4) are found in adjacent cells. This information is provided by the cellnumber values *cellx*(i), *celly*(i) and *cellx*(j), *celly*(j) accompanying these points. Thus, to detect merger, one traverses along the interface, as in the situation shown in Fig. 6.8, where the marker points along both interfaces are numbered along the clockwise direction.

Let it transpire that the points (j to j+n1) and (i to i+n2) fall in adjacent cells. This can be done by checking for, say, the point j+3 in Fig. 6.6, the four cells, i.e., top, bottom, right and left side cells. If any of these cells contains a point, then the number of that interfacial point is obtained from the value of *pointnum*. Now, however, in particular reference to point j+3, the point j+4 will lie in the bottom cell. This is merely a neighbor and not a candidate for merger with j+3. In order to avoid erroneously indicating merger in such a situation, merger is declared only under the conditions described below.

Let the index value of the interfacial point in the current cell be *pointnumber_current* and that in the adjacent cell be *pointnumber_adjacent*. *Pointnumber_current* and *pointnumber_adjacent* are obtained from the stored array, associating each cell with the corresponding interfacial marker points, namely *pointnum*.

If (an adjacent cell contains an interfacial point) then
  If (abs(*pointnumber_adjacent – pointnumber_current*) > say 4) then
      declare merger
  else
      the *pointnumber_adjacent* is just a neighbor. No merger.
  endif
endif

The latter condition is to prevent adjacent points or very close points from being declared erroneously as merger candidates. The value 4 assigned above is chosen since we expect that each cell contains no more than two interfacial markers. If merger is detected, we perform the merger operations. Thus, in the situation depicted in Fig. 6.7, one first designates the points j1 to j2 and i1 to i2 as points which have merged. Specifically, one incorporates arrays of merger points. The merger points occur in pairs, i.e., j1 connects to i1 and j2 connects to i2. Thus we declare:

$$mergept1\ (\ 1\ ) = j1$$
$$mergept2\ (\ 1\ ) = i1$$
$$mergept1\ (\ 2\ )\ = j2$$
$$mergept2\ (\ 2\ ) = i2$$

We also eliminate the points lying between i1 and i 2 and j1 and j2 since this is a region that will have merged.

Step 2. Setting up to execute merger: In the procedures above, however, assigning i1,j1 and i2, j2 as merger points and joining the pairs accordingly usually leads to very sharp curvatures in that region, since, as can be envisioned by joining j1 and i1 in Fig. 6.7, the curvature of the interface changes abruptly over one cell width. This occasions numerical difficulties associated with the large curvatures. To avoid such complications, we smooth the merger event by smearing the merger region slightly. This is done as follows: (i) Step back *nstep* points from j, i.e., j1* = j1 – *nstep* (ii) step forward *nstep* points from i1, i.e., i1* = i1+ *nstep* (iii) step back *nstep* points from i2, i.e., i 2*=i2– *nstep* (iv) step forward *nstep* points from j2 , j2* = j2+ *nstep*. The asterisk indicates the merger point after smoothing. The choice of forward or backward steps for

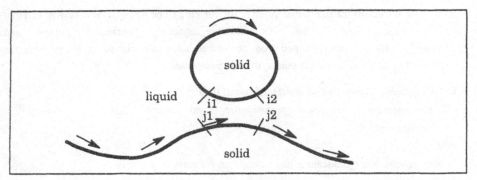

Figure 6.7. Illustration of execution of the merger procedure and interfacial point numbering.

smoothing can be directly determined by the direction of the sequence of marker points along the interface. This smoothing by *nstep* points relaxes the sharp curvature situation that may result after merger. The appropriate value of *nstep* depends on the grid spacing employed. In our calculations, for a 161 × 161 uniform grid, *nstep* = 8 has been found to be suitable, and is used in all calculations unless specified otherwise. The merger procedure may now be explained with reference to the situation illustrated in Fig. 6.8, where four circles approach each other to merge.

First, the nomenclature of interfacial point numbering is to be noted. The directions of the arrows on the interfaces shown in Fig. 6.8 indicate the direction of point numbering. The solid circles expand and merge. The merger points (after smoothing) are i1,i2, ...etc. The pairs are

mergept1 (1) = i1,   *mergept2* (1) = j1
mergept1 (2) = i2,   *mergept2* (2) = j2
mergept1 (3) = k1,   *mergept2* (3) = l1
mergept1 (4) = k2,   *mergept2* (4) = l2

and so on. Thus, the start of a merger segment is indicated by the odd-indexed merger point, while the end of the segment is given by the even-indexed merger point. Therefore, in the above, *mergept1* (1) is the starting point of a merging segment while *mergept1* (2) is the end point of that segment.

For each merging circle, we record the start point number and end point number as shown in Fig. 6.9. The string *istart* to *iend* forms a segment , i.e., one complete object. Thus for the object shown above, corresponding to the index *nseg* for example, the point number of the start

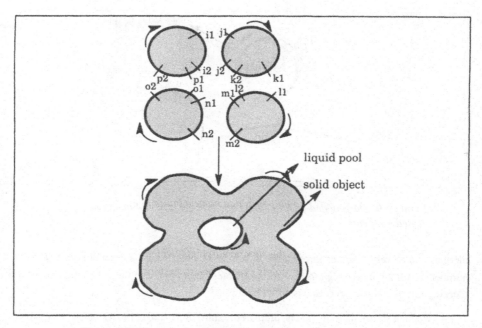

Figure 6.8. Merger of multiple objects to create more than one closed object.

point of the string defining the object is *istart (nseg)*. The end point of the string is *iend (nseg)*. Thus, each of the four circles has a designated start point and end point value. The marker points are numbered by a simple sequence, punctuated by *istart(nseg), iend (nseg)* to give the identity of each interface. Let us number the four circles 1 to 4 clockwise. Thus *istart(1)* = 1. If each circle is composed of 50 points, *iend(1)* = 51, *istart(2)* = 52, *iend(2)* = 102, etc.

Step 3. Picking up the merged interfaces: Now, after the merger has been declared to occur, one has to form a solid object as shown and one is left also with a liquid pool. The merger procedure has to pick up both these objects. Additionally, since for the interface defining the liquid pool, the solid lies to the right, the nomenclature direction for each object should be as shown in Fig. 6.9. This is so that the pool is identified correctly as being in the liquid phase. To achieve this, we proceed as follows. Let us call the start points *(istart)*, the corresponding end points *(iend)* and the merge point pairs *(mergept1, mergept2) significant* points. Such *significant* points facilitate the merger procedure by reducing the number of points to be searched to ensure merger has been completed at all candidate locations. Thus, instead of sorting through each interfacial point to determine whether all the interfacial points have been taken care of during

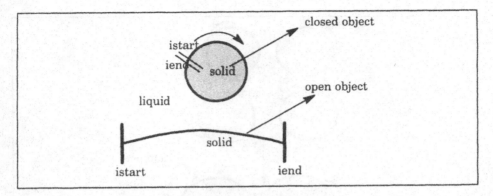

Figure 6.9. Starting and ending points and nomenclature of open and closed segments.

merger, one merely searches through the *significant* points, which are generally very few in number. Also, the use of *significant* points enables one to perform the connections between merging interfacial segments as described below.

Begin a startpoint loop, that is, until all the startpoints *istart* (1 to 4) have been covered, march along the interface collecting points without deleting them from the merger process. All the points along the interface that are encountered are immediately flagged *done* to indicate that such points have already been covered by the merger procedure. As shown in Fig. 6.8, suppose we start from *istart(1)* = 1 . We check the point lying in front of the current point along the interface. Since the next point has not yet been covered in the merger procedure, i.e. it has not been flagged *done*, we pick up that point. We proceed in this fashion along the interface, each time checking the point ahead and flagging it as done when we pick up that point. Thus, we travel clockwise on circle 1 picking up the as yet untouched points encountered. We assign the picked up points to new interfacial point arrays, say *xint_new (icount)* , *yint_new(icount)* where *icount* is a counter that records the number of points collected thus far for the new interface. We stop when we hit a *significant* point, which in this case happens to be the merge point i1. Stepping one point ahead of i1 will land us in the merger region. In this region, the points have already been eliminated from the interface. The only alternative then is to join this merge point to its pair. We thus join i1 to its corresponding merge point pair j1. We now find ourselves on circle 2. We march along that curve in the direction decided upon previously, and as shown by the arrows. If the next *significant* point encountered is an end point , which is likely, then we connect the end point to the corresponding start point (a start point for a particular segment is associated with an end point and vice versa) and carry on. If the next *significant* point is a merge point, then we check if the next point along the interface has already been eliminated from the interfacial marker array, i.e., if it is part of the merger region. If it is, then the only alternative is to connect the merge point to its pair on the other interface and continue, all the while flagging points covered as *done*. If not, then we continue to march along the same interface. The procedure results in the capture of the entire outer curve of the merged solid surface shown in the Fig. 6.8. The procedure stops when one encounters a point ahead that has already been covered (i.e. flagged *done*), which in this particular case will be *istart(1)*. At this point an entire closed curve has been traced. Thus, *istart(1)* will form the start point of the new segment, whose new *nseg* value will be 1. The *icount* value resulting after the entire shape has been traced out will be the new end point value *iend(1)* of this segment.

To pick up the remaining parts of the interfaces not covered so far, such as the pool of liquid left out in the middle, one then employs a merge point sweep, i.e., one goes to the first merge point that has not yet been covered in the previous procedure. Here this is the point i2. Beginning now from i2 (which again is promptly flagged *done*), one advances along the interface. The next *significant* point encountered is p1. Join to its pair o1 and continue. The procedure is carried out as described above and results in the pool of liquid. It is noted that the points are automatically picked up in the appropriate direction in agreement with our nomenclature convention by this procedure. All the necessary information regarding the post-merger interface is also obtained at the end of the merger treatment.

The interface undergoing merger can also be of the open type as illustrated in Fig. 6.9. Similar considerations as detailed above apply to such interfaces. The situation with the breakup of interfaces is similarly handled. For instance, consider the situation depicted in Fig. 6.10. The merge point pairs are (i1,j1) and (i2,j2). Thus, one starts at the start point of the open interval and reaches point i1. Here the points i1 to i2 have already been eliminated as merger regions. So we do not proceed beyond i1, but join it to its pair j1 and carry on. Thus the open segment has now been obtained. We then search for the leftover start points. In this case the only available start point, i.e., of the original open segment has already been covered. So a merge point sweep is performed. The merge point i2 is first encountered. One travels from i2 to j2 in the direction of the point sequence on that interface, as shown by the arrow. The region beyond j2 has been eliminated already. So one merely joins j2 to i2 and thus the second segment, of closed type, is obtained. Once a complete new segment is obtained, the type of segment is recorded, namely, whether it is open or closed. Currently, we employ and require only one open interval. The segments are arranged in the order of open intervals first, followed by closed solid and liquid pools. The merger detection procedure first takes care of the first, and only, open interface segment and then the start point and merge point searches pick up all the closed objects and left

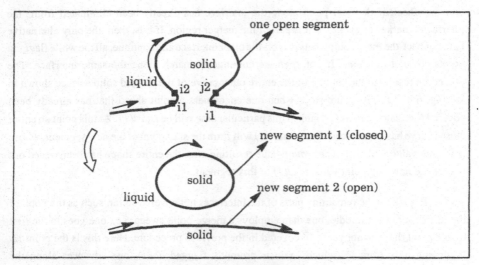

Figure 6.10. Illustration of breakup procedure.

over pools in succession. The strategy then has been to narrow the range of points to be examined to only a few significant points, thus rendering the entire merger procedure very efficient. This also adds to the logical clarity of the entire complex problem. Results obtained with the interface tracking procedure are detailed in Shyy(1994) and Udaykumar and Shyy (1994a).

The merger procedure described above may be summarized as follows:

1.   Detect merger locations.
2.   Set up data arrays.
3.   Collect open interface along with its merged appendages.
4.   Perform start point loop for collecting closed interface segments.
5.   Perform merge point loop to obtain remaining closed interfaces and pools.
6.   Redefine the interface by updating interfacial point string with newly obtained string of points.
7.   Perform the curve fit to obtain new normals and curvatures.
8.   Translate particles to new positions in cells.
9.   Perform intersection procedure to obtain fresh interfacial points.
10. Test for merger.
11. Return to step 1.

6.3   SOLUTION OF THE FIELD EQUATIONS

Now that a method has been devised to track the motion of highly curved fronts, it is necessary to compute the velocity of the front based on the computed field variable distribution and vice versa. Equation (4.2) gives the velocity of the phase front. It is evident that computation of accurate front velocities calls for accuracy in determining the temperature gradients adjacent to the front in each phase. In the case of solidification, the transport processes are governed by the Navier–Stokes equations, Eqs. (2.1a–2.1d). We solve these equations under the Boussinesq approximation (Turner 1973). In addition, the Gibbs-Thomson condition gives the temperature of the phase front as a function of the local curvature of the front. This internal boundary condition has to be incorporated in obtaining the temperature field in each phase. Furthermore, the flow variables, namely, the x and y velocity components, u and v, respectively, as well as the pressure at the interface, have to be imposed as boundary conditions in determining the flowfield. The way in which these issues are approached in formulating a pressure–based control volume discretization is described in detail in the following sections.

Consider the situation depicted in Fig. 6.11, which shows an interface lying arbitrarily on a grid representing part of the computational domain. The interfacial curve, as explained in the previous section, is tracked with the aid of marker particles indicated by crosses in the figure. Two primary tasks are to be performed in advancing the interface and thermal field in time.

1.   Given the interface position, and the boundary conditions imposed thereon, the temperature field is to be obtained at the next time level.
2.   From the gradient of the thermal field calculated, the velocities at the marker positions are computed, the interface is advanced to its new position, and the boundary condition is recomputed.

We now describe the salient features of the procedure for executing these tasks.

6.3.1   Control Volume Formulation with Moving Interface

Given the interface position, setting up to perform the flowfield calculations involves the following procedures:

(a) To classify the control volumes and associate with them the indices of the control volume markers, if any, lying on the faces on the volumes. Only two faces of the cell are allowed to be · cut by the interface.

Let us first gather all the pieces of information that knowledge of the interface location provides. Consider Fig 6.11. Joining the *interfacial markers* (x's) with straight line segments one identifies the interfacial control volumes and obtains the *control volume markers* (filled

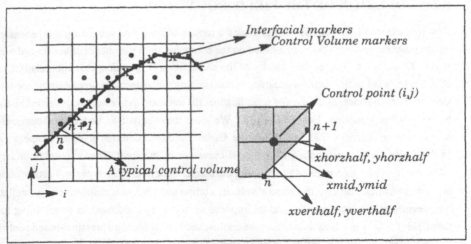

Figure 6.11. Illustration of arrangement of interfacial and control volume markers and definition of a typical control volume. n and n+1 are indices of the control volume markers.

squares) from the intersection of these segments with the grid lines. The straight line segments are consistent with the second-order accurate control volume formulation, where the field variable is assumed to vary linearly between control points. The identified interfacial cells are stored in a one-dimensional array. In each cell, the indices of the control volume (CV) markers are stored. Considering a typical control volume, as in Fig. 6.11, and knowing the x and y locations of CV markers indexed n and n+1, other necessary details can be easily extracted to gain complete knowledge of the configuration of the control volume. In particular, the intersection of the interfacial segment shown with the vertical and horizontal half cell lines are obtained. These are stored in arrays *xverthalf(i,j)*, *yverthalf(i,j)*, and *xhorzhalf(i,j)*,*yhorzhalf(i,j)*. For cells in which the horizontal or vertical half cell lines are not cut, the initial negative values assigned to these arrays are retained. Also, the midpoint of the interfacial segment is identified and designated *xmid(i,j)*,*ymid(i,j)*. Thus, each interfacial cell is fully characterized in terms of the manner in which it is cut by the interface.

(b) To identify the phase in which a control point belonging to a computational cell (indicated by the circles in Fig. 6.11) lies.

This is easily done employing the data from (a). One traverses each column of cells starting from j=1, where a flag is initialized to *solid*. Upon encountering a cell in which an interfacial segment lies, one determines whether the segment has crossed the vertical half-line

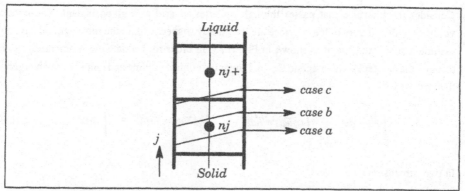

Figure 6.12. Cases in the identification of the phase of control volumes.

in that cell. Three cases are likely, as shown in Fig. 6.12. In case a, since *yverthalf(i,j)* lies below the control point (i,j), the flag is set to liquid in cell nj and the control point (i,j) is identified as lying in the liquid phase. In case b, since *yverthalf(i,j)* lies above the control point (i,j), the flag is set to liquid, while the control point (i,j) is identified as solid. This is done so that when the phase of cell nj+1 is being decided upon, the control point in that cell is identified by the flag as liquid. In case c, although the interfacial segment does lie in cell *nj*, the vertical half-cell line is not cut by the interface in cell nj. Then, the flag is retained as solid, and the identification of cell nj+1 as a liquid cell takes place in that cell. This procedure of flag resetting is repeated upon encountering an interfacial cell once more in that column.

6.3.2 The Control Volume Formulation for a Transport Variable

Based on the definition of the interface and having determined the way in which the interface lies on the underlying Cartesian grid, we are now in a position to discretize the governing equations. The discretization is performed over the control volumes defined from the procedures outlined above.

6.3.2.1 Discretization

Consider the conservation law for the variable ϕ, defined to be:
(a) unity for the continuity equation; (b) u and v for the x and y momentum equations; and (c) specific enthalpy, $C_p T$, for the energy equation. In the general case, we have

$$\frac{\partial(\varrho\phi)}{\partial t} + \nabla \cdot (\varrho\vec{u}\phi) = (\nabla \cdot \Gamma\nabla\phi) + S \tag{6.2}$$

213

Consider the interface that passes through a Cartesian grid at a given instant. The control volumes in the vicinity of the interface are irregularly shaped, and, in the most general case can assume a five-sided shape as shown in Fig. 6.13. In order to evaluate the momentum fluxes through the cell faces, we integrate Eq. (6.2) over the control volume and employ the divergence theorem to get

$$\int_V \frac{\partial \varrho \phi}{\partial t} dV + \int_A (\varrho \vec{u} \phi) \cdot \vec{n} dA = \int_A (\Gamma \nabla \phi) \cdot \vec{n} dA + \int_V S dV \qquad (6.3)$$

In two-dimensions we have

$$\int_A \frac{\partial \varrho \phi}{\partial t} dA + \oint_l (\varrho \vec{u} \phi) \cdot \vec{n} dl = \oint_l (\Gamma \nabla \phi) \cdot \vec{n} dl + \int_A S dA \qquad (6.4)$$

where the second term on the left hand side is now the line integral of the outward normal convective fluxes and the first term on the right side is the line integral of the outward normal diffusion fluxes through the faces of the control volume.

We now proceed to discretize each of the terms in Eq. (6.4) for the control volume shown. Thus

$$\int_A \frac{\partial \varrho \phi}{\partial t} dA = \frac{\varrho^{n+1}_{i,j} \phi^{n+1}_{i,j} - \varrho^n_{i,j} \phi^n_{i,j}}{\delta t} A_{cv} \qquad (6.5)$$

where the superscripts n and $n+1$ indicate the time levels and δt is the time step size. A_{cv} is the area of the irregular control volume. Next,

$$\oint_l (\varrho \vec{u} \phi) \cdot \vec{n} dl = \sum_{k=1}^{5} (\varrho_k u_k \phi_k)^{n+1} dl_k \qquad (6.6)$$

is the summation of the convective normal effluxes through each cell face of the control volume. The superscript $n+1$ indicates the implicit nature of the scheme. The diffusion fluxes are computed as

$$\oint_l (\Gamma \nabla \phi) \cdot \vec{n} dl = \sum_{k=1}^{5} (\Gamma_k (\frac{\partial \phi}{\partial n})_k)^{n+1} dl_k \qquad (6.7)$$

Let us denote the source term as follows:

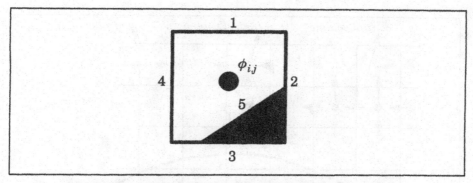

Figure 6.13. A typical control volume encountered in the mixed Eulerian–Lagrangian method.

$$\int_A S \, dA \ = \ \overline{S} \tag{6.8}$$

Substituting the above Eqs. (6.5–6.8) in Eq. (6.4), one obtains the discretized form as

$$\frac{\varrho^{n+1}_{i,j}\phi^{n+1}_{i,j} \ - \ \varrho^{n}_{i,j}\phi^{n}_{i,j}}{\delta t} A_{cv} \ + \ \sum_{k=1}^{5}(\varrho_k u_k \phi_k)^{n+1} dl_k \ = \\ \sum_{k=1}^{5}(\Gamma_k(\frac{\partial\phi}{\partial n})_k)^{n+1} dl_k \ + \ \overline{S} \tag{6.9}$$

In particular, the discrete form of the continuity equation can be written as

$$\frac{(\varrho^{n+1}_{i,j} - \varrho^{n}_{i,j})}{\delta t} A_{cv} \ + \ \sum_{k=1}^{5}(\varrho u_n)_k dl_k \ = \ 0 \tag{6.10}$$

Now, multiplying Eq. (6.10) by the value $\phi_{i,j}$ and subtracting from Eq. (6.9) gives

$$\frac{\varrho^{n}_{i,j}(\phi^{n+1}_{i,j} \ - \ \phi^{n}_{i,j})}{\delta t} A_{cv} \ + \ \sum_{k=1}^{5}\varrho_k u_k(\phi_k \ - \ \phi^{n+1}_{i,j})^{n+1} dl_k \ = \\ \sum_{k=1}^{5}(\Gamma_k(\frac{\partial\phi}{\partial n})_k)^{n+1} dl_k \ + \ \overline{S} \tag{6.11}$$

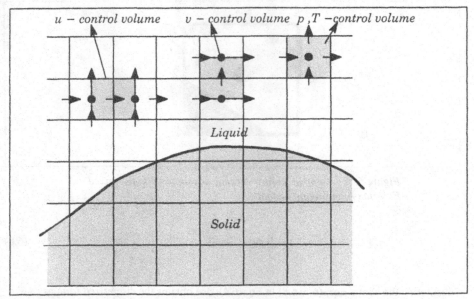

*Figure 6.14. Control volumes for u,v,p, and T variables showing the
staggered arrangement*

6.3.2.2   Treatment of Variables on the Staggered Grid

The solution of the discrete form, namely Eq. (6.11), is carried out on a staggered grid
arrangement shown in Fig. 6.14. Staggered grids have been extensively adopted for computation
of incompressible fluid flows due to their many advantages (Patankar 1980, Shyy 1994). The
variables and their respective control points are located as indicated in Fig. 6.14. The interface
is thus required to be tracked on such a grid arrangement. The definition of the control volumes,
which is carried out according to the intersection procedure, is now applicable to each type of
control volume shown, namely the u, v, and (p,T) control volumes. For a given control volume,
then, the interface tracking procedure provides the information shown in Fig. 6.15. Thus, at each
iteration, explicit definition of the interface location/control volumes for each variable and its
grid is available. This facilitates the application of boundary conditions on the faces of the cells
containing the interface, as explained later.

6.3.2.3   Computation of Convective Fluxes

We may return now to Eq. (6.11), for the situation shown in Fig. 6.15 below. It is standard in
the notation of pressure-based implicit methods to identify the variables at time level n with
superscript o and leave the variables at the current time level $n+1$ without superscripts. Using

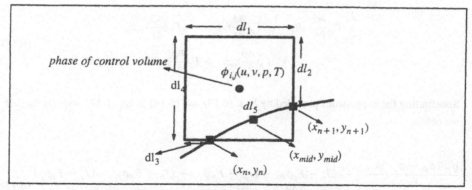

Figure 6.15. Illustration of information provided by the interface tracking module during assembly of coefficients for control volumes.

this conventional notation and rewriting Eq. (6.11) for the variable ϕ, the convective fluxes are evaluated as

$$\sum_{k=1}^{5} (\varrho u_n)_k(\phi_k - \phi_P)dl_k = (\varrho_e u_e)(\phi_E - \phi_P)dl_e - (\varrho_w u_w)(\phi_W - \phi_P)dl_w$$
$$+ (\varrho_n u_n)(\phi_N - \phi_P)dl_n - (\varrho_s u_s)(\phi_S - \phi_P)dl_s \pm (\varrho_I u_{nI})(\phi_I - \phi_P)dl_I$$

$$(6.12)$$

which may be written as

$$\sum_{k=1}^{5} (\varrho u_n)_k(\phi_k - \phi_P)dl_k = F_e(\phi_e - \phi_P) - F_w(\phi_w - \phi_P)$$
$$+ F_n(\phi_n - \phi_P) - F_s(\phi_s - \phi_P) \pm \varrho_I u_{nI}(\phi_I - \phi_P)dl_I$$

$$(6.13)$$

Here the F's stand for the mass fluxes through the faces of the control volume. The lower case subscripts indicate the values at the cell faces. Subscript I indicates the values at the interface. The manner in which these fluxes are evaluated , i.e., the specific shape function assumed for the variable ϕ for evaluation of these fluxes determines the order of accuracy of the scheme employed. In Eq. (6.13), the sign of the last term is to be decided. What is required is the outward normal flux from the interfacial segment. We shall return to this point momentarily.

6.3.2.4 Evaluation of the Diffusion Fluxes and the Full Discretized Form

Now, the diffusion fluxes can be written as

$$\sum_{k=1}^{5} \Gamma_k(\frac{\partial\phi}{\partial n})_k dl_k = \Gamma_e(\frac{\partial\phi}{\partial x})_e dl_e - \Gamma_w(\frac{\partial\phi}{\partial x})_w dl_w$$
$$+ \Gamma_n(\frac{\partial\phi}{\partial y})_n dl_n - \Gamma_s(\frac{\partial\phi}{\partial y})_s dl_s \pm \Gamma_I(\frac{\partial\phi}{\partial n})_I dl_I \qquad (6.14)$$

Substituting the expressions provided by Eqs. (6.13) and (6.14) in Eq. (6.11) and simplifying, we obtain

$$\frac{\varrho_P{}^o(\phi_P - \phi_P{}^o)A_{cv}}{\delta t} + (J_e - F_e\phi_P) - (J_w - F_w\phi_P) + (J_n - F_n\phi_P) - (J_s - F_s\phi_P)$$
$$= \pm \Gamma_I(\frac{\partial\phi}{\partial n})_I dl_I \mp \varrho_I u_{nI} dl_I(\phi_I - \phi_P) + \overline{S} \qquad (6.15)$$

where

$$J_e = \varrho_e u_e dl_e \phi_e - \Gamma_e(\frac{\partial\phi}{\partial x})_e dl_e \qquad (6.16)$$

$$J_n = \varrho_n v_n dl_n \phi_n - \Gamma_n(\frac{\partial\phi}{\partial y})_n dl_n \qquad (6.17)$$

with similar expressions for J_w and J_s. The following notation is standard for the SIMPLE algorithm (Patankar 1980),

$$J_e - F_e\phi_P = a_E(\phi_P - \phi_E) \qquad (6.18)$$

$$J_w - F_w\phi_P = a_W(\phi_W - \phi_P) \qquad (6.19)$$

with similar expressions for a_N and a_S. The forms assumed by $a_{E,W,N,S}$, along with the source term treatment, determine the order of accuracy of the differencing scheme employed. For example,

$$a_E = D_e A(|P_e|) + [-F_e, 0] \qquad (6.20)$$

where P_e is the Peclet number ($= F_e/D_e$) and the square bracket implies the maximum of the two quantities. In this work, the second-order central difference scheme is employed to discretize spatial derivatives, for which

$$A(|P|) = 1 - 0.5|P| \qquad (6.21)$$

Equation (6.15) can now be written as

$$\frac{\varrho_P{}^o(\phi_P - \phi_P{}^o)A_{cv}}{\delta t} + a_E(\phi_P - \phi_E) - a_W(\phi_W - \phi_P) + a_N(\phi_P - \phi_N) - a_S(\phi_S - \phi_P)$$

$$= \pm \Gamma_I(\frac{\partial \phi}{\partial n})_I dl_I \mp \varrho_I u_{n_I} dl_I(\phi_I - \phi_P) + \overline{S} \tag{6.22}$$

Letting

$$\frac{\varrho_P{}^o A_{cv}}{\delta t} = a_P{}^o \tag{6.23}$$

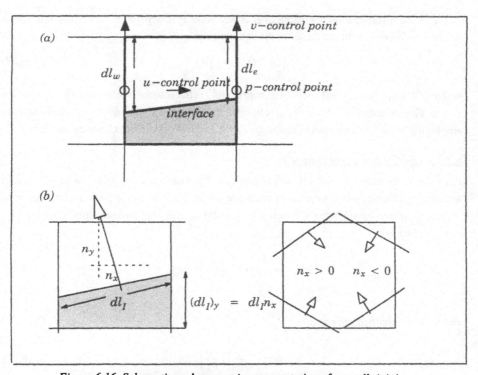

Figure 6.16. Schematic and geometric representation of cut cell. (a) A typical control volume with an interfacial segment. (b) Directions of normals and projected lengths for a typical control volume. n_x and n_y are components of unit normal in the x and y directions.

and

$$a_P = a_P{}^o + a_E + a_W + a_N + a_S \tag{6.24}$$

we have the final discretized form

$$a_P\phi_P = a_N\phi_N + a_S\phi_S + a_E\phi_E + a_W\phi_W + b \tag{6.25}$$

where

$$b = \bar{S} \pm \Gamma_I(\frac{\partial\phi}{\partial n_I})dl_I \mp \varrho_I u_{nI} dl_I(\phi_I - \phi_P) \tag{6.26}$$

In the above discretization, the value of ϕ at any point P is dependent on its four immediate neighbors. When cast in matrix form, the equation reads

$$[C] \ \{\Phi\} = \{B\} \tag{6.27}$$

where $[C]$ is a pentadiagonal coefficient matrix, $\{\Phi\}$ is the solution vector, and $\{B\}$ is a source vector. The procedure usually adopted is to solve the system of equations in an iterative fashion employing a line SOR method in conjunction with a tridiagonal matrix solution procedure.

6.3.2.5 *Evaluation of the Source Term*

For each of the equations, namely the continuity and momentum equations, the source term $\bar{S}$ assumes a different form. The pressure terms are included in this source term and its evaluation involves certain considerations which we now detail. For the u-momentum equation, for example, we have

$$\bar{S}_u = \int_A - \left(\frac{\partial p}{\partial x}\right) dA \tag{6.28}$$

and for the v-momentum equation

$$\bar{S}_v = \int_A - \left(\frac{\partial p}{\partial y}\right) dA \tag{6.29}$$

Consider the term corresponding to the u-momentum equation, as in Eq. (6.28), in relation to the u-control volume, shown in Fig. 6.16(a). Applying the divergence theorem, Eq. (6.28) can be written as

$$\overline{S}_u \;=\; -\sum_{k=1}^{5} \, p_i (dl_y)_k \tag{6.30}$$

where the length dl_y is the projected length of a side along the y-direction. Thus, for the control volume shown, the pressure contributions take the form

$$\overline{S}_u \;=\; -\, p_e dl_e \;+\; p_w dl_w \;\pm\; p_f (dl_I)_y \tag{6.31}$$

Now, as shown in Fig. 6.16(b), $(dl_I)_y = dl_I \, n_x$, where n_x is the x-component of the unit normal vector to the interface in that cell, and is already available from the interface tracking information. From observing Fig. 6.16(b), it is evident that for $n_x < 0$ a negative sign is required in front of the pressure term for the interfacial segment and for $n_x > 0$, a positive sign is required. Thus the appropriate form for the source term is

$$\overline{S}_u \;=\; -\, p_e dl_e \;+\; p_w dl_w \;+\; p_f dl_I n_x \tag{6.32}$$

To estimate the value of pressure at the interface, p_I, a bilinear extrapolation is performed from the neighboring liquid phase pressure control points onto the interface location.

6.3.2.6 *Computation of Interfacial Fluxes*

We now proceed to detail the method for obtaining the interfacial flux terms. Considering the fluxes from the interface,

$$\pm \, \Gamma_f (\frac{\partial \phi}{\partial n})_f dl_I \;\mp\; \varrho_f u_{nf} dl_I (\phi_I - \phi_P) \tag{6.33}$$

We need to estimate the following quantities at the interface;

$$(\frac{\partial \phi}{\partial n})_I \;,\;\; \varrho_f u_{nI} \;,\;\; \phi_I \tag{6.34}$$

and determine the missing signs to evaluate the flux terms. As for the above values at the interface, ϱ_I is easily obtained for incompressible flow as the value of the density at the liquid control point closest to the interface. u_{nI}, the fluid velocity normal to the interface is evaluated from the boundary condition at the interface, once the interfacial velocity at that point is known. The interfacial velocity is, of course, determined from the Stefan condition, Eq. (4.2), during the course of the calculation and is coupled to the temperature gradients in the vicinity of the interface. The normal gradient of any variable ϕ at the interface, i.e., $(\partial \phi / \partial n)_I$ is evaluated in the same fashion as for the pure conduction problem. A probe is inserted in each phase and a biquadratic shape function is described in the vicinity of the interface. The derivative can then

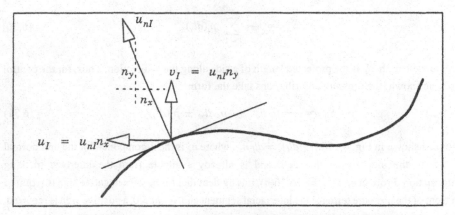

Figure 6.17. Illustration of velocity components of fluid at the interface.

be estimated. The value of the variable on the interface, ϕ_I, is again obtained from the boundary conditions. In particular, for:

1. The u-momentum equation $\phi = u$ and $u_I = u_{nI}\, n_x$, as shown in Fig. 6.17.
2. The v-momentum equation $\phi = v$ and $v_I = v_{nI}\, n_y$, as shown in Fig. 6.17.
3. For the energy equation $\phi = T$ and T_I is obtained from the Gibbs-Thomson condition at the interface.

Next we need to decide upon the signs to be assigned to each of the interfacial fluxes. The diffusion flux with the as yet undetermined sign is

$$\pm\ \Gamma_I(\frac{\partial\phi}{\partial n})_I dl_I \tag{6.35}$$

The positive sign is applicable when an outflux is evaluated.

We now detail the procedure to obtain the value of the gradient at the interface $(\frac{\partial\phi}{\partial n})_I$ with the temperature variable Θ as an example. Consider again the situation shown in Fig. 6.11. As a first step, the control volume formulation is to be generalized to the polygonal (maximum 5-sided) control volumes that occur at the interface. Interfacial features requiring finer grid resolution will be automatically smoothed out, in accordance with the merger procedure developed. The flux through the interfacial segment, side 5, is crucial as far as interfacial behavior is concerned. Before we proceed with the description of the method to evaluate $(\frac{\partial\Theta}{\partial n})_5$, we call attention to the interfacial normal velocity expression, Eq. (4.2). The calculation of

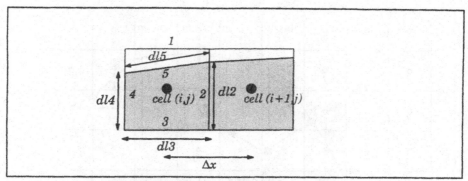

Figure 6.18. Nomenclature of control volume sides and dimensions.

$(\frac{\partial \Theta}{\partial n})_5$ thus serves two purposes. First, it supplies the interfacial heat flux to the temperature field calculation. Second, this quantity will be required in computing the interfacial velocity. The fact that the same procedure is used for both purposes renders the interface propagation computations consistent with the field solver.

Now consider the cell (i,j) shown in Fig 6.18. In order to obtain the normal gradient at the interfacial segment in that cell, $(\frac{\partial \Theta}{\partial n})_5$, we project a probe into each phase in the direction of the normal to the interface at that segment, as shown in Fig. 6.19 and Fig. 6.20. Let the normal be given by $\vec{n} = n_x \vec{i} + n_y \vec{j}$. Let the length of the probe be dn. This length is chosen so that

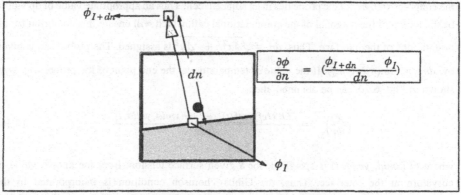

$$\frac{\partial \phi}{\partial n} = (\frac{\phi_{I+dn} - \phi_I}{dn})$$

Figure 6.19. Interfacial gradient of a field variable obtained by the normal projection procedure.

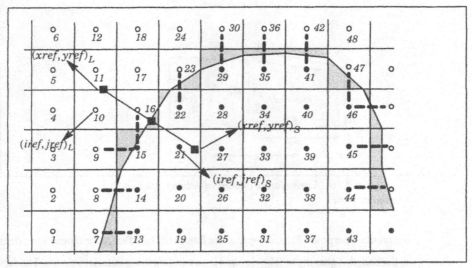

Figure 6.20. Illustration of cut cells, definition of partner cells, and method
for evaluating normal gradient at the interface (see cell 16). Normal is
projected from point (xmid,ymid) in cell 16 along the normal at that point.
The probe of length dp ends at the points (xref, yref) in each phase.
(iref,jref) is the index of cell in which (xref,yref) lies. Solid phase control
points are denoted by filled circles and liquid phase by open circles.
Partner cells are indicated by the linkages between cells.

the endpoint *xref,yref* of the probe lies in an adjacent cell. Thus an appropriate value of dp would
be the length of the diagonal of the computational cell, which will ensure this condition for all
orientations of the interface. Thus, $dn = \sqrt{\Delta x^2 + \Delta y^2}$ is assigned. The probe then is given
by, $\vec{dn} = dn(n_x\vec{i} + n_y\vec{j})$. If the value of temperature at the end point of the probe, *xref, yref,*
shown in Fig. 6.20, can be obtained, then,

$$\left(\frac{\partial \Theta}{\partial n}\right)_s = \frac{\Theta(xref, yref, t) - \Theta(xmid, ymid, t)}{dn} \qquad (6.36)$$

where Θ *(xmid, ymid, t)* is specified, for a given surface tension, from the known value of
curvature at the interface. Thus, the Gibbs–Thomson condition is incorporated in the
calculation. The locations *xref=xmid*(i,j) + *dn* n_x and *yref=ymid*(i,j) + *dn* n_y. The cell (iref,jref)
in which point *(xref,yref)* lies is easily identified. The phase in which *(xref,yref)* lies is known
a priori, since, by convention, the normal is specified to point from the solid to liquid. The values

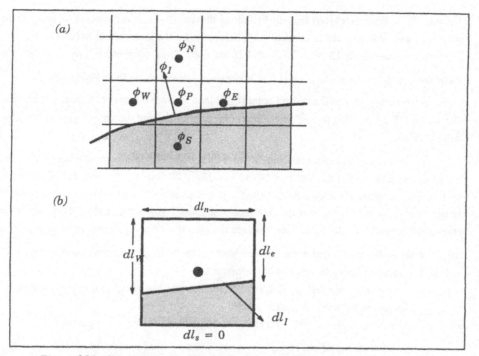

Figure 6.21. Illustration of nomenclature. (a) control points, and (b) control volume sides.

of normal gradients in each phase are, therefore, extracted by projecting probes into both phases along the normal direction.

Now, to obtain the value of temperature $\Theta(xref, yref, t)$, a biquadratic shape function is fit to the temperature field in the vicinity of the interface. The six coefficients appearing in the biquadratic form, $F(x,y) = ax^2 + bxy + cy^2 + dx + ey + f$, are obtained by choosing six cells in the *same phase* around the point *(xref, yref)* and inverting the resulting 6 x 6 matrix by a Gaussian elimination procedure. In this process, it is also to be ensured that the six points chosen for fitting the biquadratic should cover three i and three j levels, which is obviously necessary for a biquadratic representation to be valid. For example, in the case of the cell 16, the probe, extended from the point *(xmid, ymid)* in that cell, ends in cell 10. In order to obtain the value of temperature at the point *(xref, yref)* which lies in the liquid phase, we need to choose six cells in the liquid phase. The choice of five cells is straightforward in this case. These are cells numbered 4,16,11,9, and 10 itself. The sixth cell remains to be chosen. Depending on which quadrant of the cell 10 the point *(xref,yref)* lies, the cell along the diagonal that is closest is picked. In this particular example, that cell is numbered 17. Similarly, for the temperature at *(xref,yref)*s, i.e., in the solid phase, cells 15,21,27,22,20 and 26 are used. Thus, once $\Theta(xref,yref,t)$ is obtained from the functional form, the normal flux is obtained in each phase from Eq. (6.36). The $\frac{\partial\Theta}{\partial n}$ value in the *same phase* as the control point (i,j) is then fed into the control volume expression for that cell. Thus, the flux through each face of the control volume shown in Fig. 6.21 has now been obtained.

To ensure conservation in the vicinity of the interface, however, several other procedures are necessary. Consider cells numbered 29 and 30 in Fig. 6.20. Between the two control volumes, there is a trapezoidal piece, shown hatched, that is unaccounted for. To enforce conservation, it is necessary that the flux across the interface computed for cell 29, as described above, be transmitted to cell 30. This procedure involves the identification of three types of cells:

a) *Bulk cells.* A cell that has all its neighbors in the same phase. The control volume formulation for such cells is straightforward.

b) *Interface adjoining cell.* A cell, such as cell 30, that has a neighbor in the opposite phase, but is not itself an interfacial cell.

c) *An interfacial cell.* A cell, such as 29, through which the interface passes.

To enforce conservation, for cells of type b and c we need to identify *partner cells.* The choice of partner cells has been indicated in Fig. 6.20 for each cell by the linkages shown. Each interfacial cell is assigned one or more partner cells (for example, cells numbered 9,15,16). When the control volume fluxes are assembled for the interfacial cells, the fluxes into the designated partner cells are modified appropriately. For example, let cell 29 belong to the solid phase, and cell 30 lie in the liquid. While performing the flux calculations for cell 29, we compute the $\frac{\partial\Theta}{\partial n}$ value in each phase at the interface. The flux $\frac{\partial\Theta}{\partial n}$ value in the liquid phase is then transmitted to cell 30. We recall that these fluxes also go into determining the normal velocity of the interface thereby maintaining consistency. In addition, the control volume shape and area of the partner cell are redefined, so that cell 30, for instance, absorbs the hatched trapezoidal region. Thus, there are no missing pieces in the mosaic of control volumes and all the hatched regions are incorporated into the appropriate partner cells, along with the fluxes at the interface. This assembly process is not difficult to execute. The procedures only apply to the one-dimensional interfacial cell array, and there are only eight separate cases of interfacial cell and partner types, as can be deduced from Fig. 6.20.

The control volume flux assembly procedures have been checked for symmetry by examining the numerical values of the difference coefficients for a specified symmetric interface shape and letting it evolve over time. Also, the cell assembly direction was reversed, so that the interfacial cell and its partner exchanged roles, which yielded identical results.

By the procedure outlined above, the gradient for any variable ϕ is evaluated as follows:

$$\frac{\partial \phi}{\partial n} = (\frac{\phi_{I+dn} - \phi_I}{dn})$$ (6.37)

for the situation shown in Fig. 6.19.

The gradient as evaluated above represents the influx into the control volume in each phase. A negative sign is, therefore, required to provide the efflux from the control volume. As for the sign of the convective interfacial flux, we have adopted the convention that the normal points into the liquid phase. Hence u_{nI} is the normal velocity of the fluid at the interface, and is directed into the liquid phase. Therefore, the convective flux into the control volume is given by

$$\varrho_I u_{nI} dl_I (\phi_I - \phi_P)$$ (6.38)

To obtain an efflux from the control volume, a negative sign should appear in front of this flux term. Thus the final form of the source term b is given by

$$b = -\Gamma_I dl_I (\frac{\partial \phi}{\partial n})_I + \varrho_I u_{nI} dl_I (\phi_I - \phi_P) + \overline{S}$$ (6.39)

6.3.2.7 Computation of the Pressure Field

For incompressible flows, since no explicit equation exists for the pressure field, some method needs to be devised to compute the pressure field. To extract the pressure, one makes use of the continuity equation and obtains a correction equation for the pressure and velocities that enforces mass conservation. The correction procedure is continued until convergence is achieved for each time step. Such a pressure correction is described in detail in Chapter 2. The pressure correction at each point is computed from

$$a_P p'_P - a_N p'_N - a_S p'_S - a_E p'_E - a_W p'_W =$$
$$\frac{(\varrho_P^o - \varrho_P)A_{cv}}{\delta t} + \varrho_w u_w dl_w - \varrho_e u_e dl_e + \varrho_s v_s dl_s - \varrho_n v_n dl_n$$ (6.40)

where the superscript 0 denotes the value at the previous time step, and p' is the pressure correction at each outer iteration at the current time step. The right hand side represents the mass deficit in the control cell, and is required to be nullified at convergence. The individual terms on the right hand side are mass fluxes through the faces of the control volume, including the

227

interfacial segments. Such terms are easily evaluated, since the velocity values required in computing the mass fluxes are available at the cell faces, a feature of the staggered grid. The coefficients for the pressure correction equation are assembled in the standard way (Patankar 1980).

6.3.2.8 Computing the Velocities of the Interfacial Markers

At this stage we have already obtained the information necessary for computing the normal velocities of the interface. The gradients $(\frac{\partial \Theta}{\partial n})_{l,s}$ are now available at the locations (xmid, ymid) in each control volume. However, the interfacial markers are not usually located at the points (xmid, ymid). Thus, one has to obtain the value of the temperature gradients at the locations of these markers in order to calculate their velocities. To transmit this information to the locations of the interfacial markers, we utilize the array corresponding to the locations (xmid, ymid) along the interface. The value of the arclength at each point (xmid, ymid) while traversing the interface is obtained and the $(\frac{\partial \Theta}{\partial n})_{l,s}$ values are fit to piecewise quadratic functions of arclength, i.e.,

$$(\frac{\partial \Theta}{\partial n})_{l,s} = a_i \psi^2 + b_i \psi + c_i \tag{6.41}$$

where ψ is the arclength along the interface in the ith interface segment. Thus, knowing the value of ψ corresponding to the interfacial marker locations, one can identify the segment in which the point lies and the value of $(\frac{\partial \Theta}{\partial n})_{l,s}$ can be calculated from the appropriate functional form. The interfacial marker is then translated along the normal $\vec{n}$ such that

$$x^{n+1} = x^n + \Delta t v_n n_x \ , \ y^{n+1} = y^n + \Delta t v_n n_y \tag{6.42}$$

This specifies the new location of the interfacial marker at convergence.

6.3.2.9 Dealing with Cut Cells

When the interface passes through the staggered grid arrangement, as shown in Fig. 6.15, the control volumes in the vicinity of the interface become fragmented. Consider the control volume for the u-velocity $u(i,j)$, as shown in Fig. 6.22(a). The u control point lies in the liquid phase here while a piece of the control volume, shown shaded, lies in the solid phase. However, in Fig. 6.22(b), the shaded region belongs to cell (i,j−1) but lies in the solid phase. Thus, one has to account for the existence of such pieces of control volumes and assign them to the proper phase. This is done by defining partner cells for each type of cell, a procedure discussed in detail above. In Fig. 6.22(a), for example, the cell (i,j) is now redefined to be the unshaded irregular cell. Similarly, in Fig. 6.22(b) the new u-cell (i,j) is defined by the shaded region. This procedure of redefining the cut cells and their partners is accomplished by running through the array of

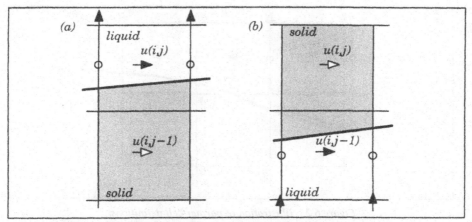

Figure 6.22. Definition of cut cells and their partners. (a) Present cell is in liquid phase. (b) Present cell is in solid phase.

interfacial cells. By assigning partner cell fluxes and redefining dimensions, it is possible to maintain consistency at cell faces and conservation of fluxes.

6.3.2.10 Conservation and Consistency at Cell Faces

In solving the set of conservation laws, it is important to set up the control volume formulation and differencing schemes so that strict conservation of fluxes is maintained. Consider the situation shown in Fig. 6.23. Of concern to us are the flux evaluation through the faces of the cells (i,j) and (i $\pm$ 1, j) and their relation to the partner cell definitions. The newly redefined cells, after partner cell assignment, are as shown by the dotted lines. The fluxes through the west face of (i,j) are now magnified by the factor $(dl_w(i,j) + (dy - dl_w(i,j-1))) / dl_w(i,j)$ to correct for the augmented length of the west face. A similar modification is made for cell (i–1, j) when its partner cell (i–1, j–1) is being dealt with. Similarly, in cell (i,j–1) the pressure force on the west side is magnified by the factor above. Fluxes from the interface computed in cell (i, j–1) are assigned to cell (i,j) and from cell (i–1,j–1) to cell (i–1,j). Thus, the partner assignment procedure and the cut cell coefficient assembly algorithms achieve explicit flux conservation in the cells affected by the interface.

6.3.2.11 Anomalous Cases

Some anomalous situations may be encountered in the coefficient assembly process. For instance, in Fig. 6.24, the u-control point (i,j) is adjoined by a solid phase control point. Therefore, the flux through the east face is not straightforward to compute. The way this is done to maintain consistency at that face is by noting that cell (i+1,j+1) has been redefined by the

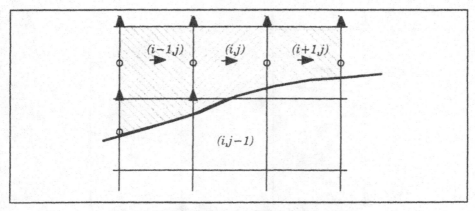

Figure 6.23. Illustration of partner cell definitions.

partner cell assignment procedure to be of the shape shown. Thus a part of the west face of cell (i+1,j+1) is now adjacent to the east face of cell (i,j). Therefore, the flux through the west face of cell (i+1,j+1) has to be consistent with the fluxes through the east faces of (i,j) and (i,j+1). The flux through the east face of cell (i,j) is now obtained by redistributing the west face flux of cell (i+1,j+1), weighted according to the dimensions $dl_e(i,j)$ and $dl_e(i,j+1)$. In addition, if the east face pressure control point of cell (i,j) is also in the solid phase, the pressure contribution from cell (i+1,j+1) is redistributed consistently between cells (i,j+1) and (i,j). Such anomalous cases exist for all eight types of cells and for all control volumes. Treating such cases adds to the tedium of assembly of the coefficients but is critical in obtaining solutions to flow problems.

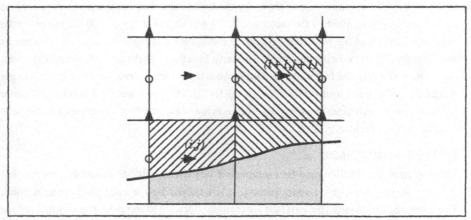

Figure 6.24. Illustration of an anomalous case in defining fluxes.

If consistency at cell faces is violated, convergence cannot be achieved due to the existence of the spurious sources/sinks of mass, momentum, and energy at such locations.

6.3.2.12 Distinction Between Liquid and Solid Cells

In the cell assembly procedure, one has to be careful not to step across the interface into the opposite phase in performing the flux computations. The advantage in combining the Eulerian (field solver) and Lagrangian (interface tracking) methods here is that the two phases can be treated separately. Thus, unlike in the Eulerian methods, the interface separating the two phases can be explicitly defined and treated as a discontinuity. And, in contrast with the Lagrangian methods, grid redistribution for conformity with the moving, distorting boundary is avoided. Thus, in each phase the operations involved in obtaining the flux estimates should involve points in the same phase and the interfacial values only. The treatment of the liquid and solid phase control points follows the same procedure as far as the coefficient assembly is concerned. However, the solid is passive as far as fluid flow goes, unlike in the purely Eulerian methods where it is assumed to have some porosity in the proximity of the interface to ease the sudden property jump across it. In our case, once the coefficients are assembled, a flag denoting the solid is employed to turn off the fluid flow computations in the solid phase, such that $a_P = 1$ and $a_E = a_W = a_N = a_S = b = 0$ in the solid, which results in $u = v = p' = 0$ in the solid phase. This, of course, is not applicable for the energy equation where heat conduction in the solid is to be accounted for.

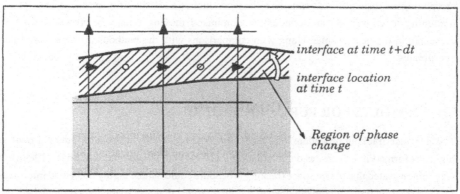

Figure 6.25. Change of phase of cells as the interface traverses the domain.

6.3.2.13 *Moving Boundary Problems – Treatment of Cells that Change Phase*

When the interface moves, say, due to change of phase, the shape of the interface at two time instants (or iterations) may develop as shown in Fig. 6.25. Thus, cells (i,j), (i+1,j), etc., have moved from the solid phase into the liquid phase. In the implicit solution procedure adopted here, information regarding the previous state of a control point is required to evaluate the time derivative, that is, the quantity $\phi^n_{i,j}$, corresponding to the new phase, is non-existent for that control point. To overcome this difficulty, flags are employed to indicate the current and previous phases of the control point. When a control point changes phase, the value at that location, corresponding to the phase in which it finds itself, is estimated by performing a linear interpolation, as shown in Fig. 6.26. The new value in a cell which undergoes phase change is thus obtained by interpolation from neighboring cells in the same phase as

$$\phi^n_{i,j} = (dy_I \phi^n_{i,j+1} + dy_N \phi^n_I)/(dy_I + dy_N) \qquad (6.43)$$

In particular, the value $\phi^n_{i,j}$ is then obtained via the implicit solution procedure by iteration. In our computations, flipping of a control point between two phases was practically non-existent. This problem exists in purely Eulerian methods. In our case, since the interface is a continuous entity, each point on the interface is influenced by the motion of all its neighbors along the interfacial curve and this global influence helps in avoiding the flipping of phase. The interface reconstruction procedures rely on local fluid fraction information in a cell to define the interface and, thus, it is possible for a cell to flip between phases in the course of the iterations. Extensive tests have been performed to assess the performance of ELAFINT. In the following, morphological instabilities dominated by heat conduction and capillarity are simulated. Results pertaining to convection and conduction dominated moving boundary problems will be presented in the next chapter, along with comparisons with the methods utilizing curvilinear coordinates.

6.4   RESULTS FOR PURE CONDUCTION

The interface tracking procedure was tested, employing velocity models to induce growth of highly deformed interfaces, as detailed in Shyy (1994) and Udaykumar and Shyy (1994a). It was demonstrated that the interface tracking procedure could handle highly deformed interfaces. The intersection procedure maintained a smooth interface that was adequately replenished with markers as the interface deformed. Mergers and breakups and a combination of these events were successfully treated. Quantities pertaining to the interfacial curve, such as slopes and curvatures, were obtained accurately. Thus, the interface tracking algorithm, while explicitly tracking the

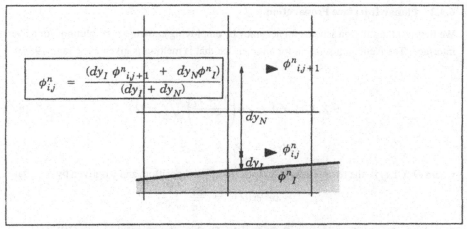

$$\phi_{i,j}^n = \frac{(dy_I \, \phi_{i,j+1}^n + dy_N \phi_I^n)}{(dy_I + dy_N)}$$

Figure 6.26. Illustration of interpolation procedure for the estimation of field quantities in cells that have changed phase.

interfacial curve, overcomes the conventional drawbacks of a purely Lagrangian scheme. As the next step, we compute the growth of an unstable solidifying front with pure conduction heat transport in both phases. The pure conduction approximation is frequently adopted in the analysis of microstructural evolution, such as in diffusion–controlled dendritic or cellular growth. The interface is driven by the difference in thermal gradients between the two phases and is computed from the Stefan condition.

6.4.1   Grid Addition/Deletion

The computational domain is configured as shown in Fig. 6.27. The domain is partitioned into three regions. Coarse grids are employed in the regions I and III, away from the interface, while fine grids are employed in the region II, close to the interface. However, as the interface rapidly grows out of region II there is a need for introducing a fine grid that precedes the interface in order to be able to calculate the gradients ahead of it with desired accuracy. To achieve this, grid lines are added such that a sufficiently extended region ahead of the interface is replenished with fine grids throughout the evolution of the interface. The values of field variables in this region are obtained by linear interpolation. The grid addition takes place at a frequency depending on the extent of the domain traversed by the interface.

6.4.2 Planar Interface Propagation

We first test the solution scheme for accuracy by employing known, exact solutions for a planar interface. The Neumann solution for an interface that is melting is given by (Crank 1984):

$$\Theta(X, Y, \tau) = 1 - \text{erf}\left(\frac{Y - Y_l}{2\sqrt{\tau}}\right) \Bigg/ \text{erf}(\lambda) \tag{6.44}$$

$$S(\tau) = Y_l - 2\lambda\sqrt{\tau} \tag{6.45}$$

where $\Theta(X, Y, \tau)$ is the temperature, $S(\tau)$ is the interfacial position, and ξ is given by the relation

$$\lambda e^{\lambda^2}\text{erf}(\lambda) = \frac{St}{\sqrt{\pi}} \tag{6.46}$$

St is the Stefan number given by $C_p(T_l - T_m)/L$. The boundary conditions are $\Theta(X, Y_l, \tau) = 1$ at the liquid side and $\Theta(X, Y, \tau) = 0$, $Y \leq S(\tau)$, $\tau \geq 0$, i.e., the solid is at uniform temperature. The initial conditions are specified at time $\tau_o > 0$, taking account of the singularity at $\tau = 0$ (Yao and Prusa, 1989). Thus

$$\Theta(X, Y, \tau_o) = 1 - \text{erf}\left(\frac{Y - Y_l}{2\sqrt{\tau_o}}\right) \Bigg/ \text{erf}(\lambda) \tag{6.47}$$

and the initial interface location is given by the expression

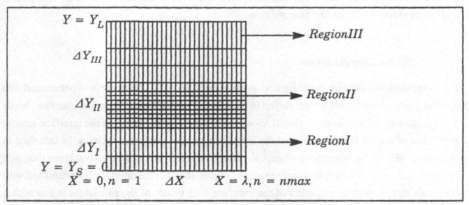

Figure 6.27. Grid arrangement for the computational domain.
Region II extends along with the interface

$$S(\tau_o) = 2\lambda \sqrt{\tau_o} \tag{6.48}$$

This test problem has been used in Chapters 4 and 5. Here we test the performance of ELAFINT with several Stefan numbers.

Figure 6.28 compares the exact and numerical solutions obtained for Stefan number St= 0.1303, λ=0.25. The calculations were performed with 41 grids along X and 41 grids along Y direction. The value of Y_l was taken to be 4. The initial position of the interface was at $Ymean$=3.61. The computation was performed for 1000 time steps, and over the period of calculation, the computed and exact interface locations are in excellent agreement, as seen in Fig. 6.28(a). The superposed exact and computed temperature fields are shown in Fig. 6.28(b).

A more stringent test of the planar interface propagation is in computing the higher Stefan number case, St=2.8576, λ=0.9, shown in Fig. 6.29. The computations are performed with 61 grid points along the y-direction. Again, remarkable agreement is obtained for computations over 100 time steps. For this higher Stefan number case, the interface moves more rapidly and a significant portion of the domain (Y_l = 4) has been traversed by the interface.

From the above experiments, it is clear that for a planar interface, the conservation statement across the interface is honored by the control volume formulation and the interface position is accurately obtained for a wide range of Stefan numbers.

6.4.3 Non-planar Interfaces

In the computations to follow, we employ *nmax* grid points along the *x*–direction, where *nmax* is 21, 41 or 81. Since χ=4, and Y_L = 40 in all subsequent calculations, we have ΔX=4/*nmax*. Depending on the value of Δx, we employ grid points in the region II, such that $\Delta Y_{II} = \Delta X$. As mentioned before, as the interface grows and propagates, grids are added ahead of, and deleted behind, the interface. Thus the extent of region II increases as the computation proceeds.

The domain of computation is shown in Fig. 6.30. The liquid end was maintained at Θ_l = –40, with the solid side at Θ = 1. The side boundaries are adiabatic, i.e. $\frac{\partial \Theta}{\partial x}$ = 0. The system of equations solved and the boundary conditions applied, render the situation more akin to viscous fingering in Hele–Shaw cells than to a free dendritic growth environment. Thus, comparison with the boundary integral computations (DeGregoria and Schwartz 1986) of that problem is justified, although such computations are performed for a semi–infinite domain. Detailed discussions can be found in Udaykumar and Shyy (1994b).

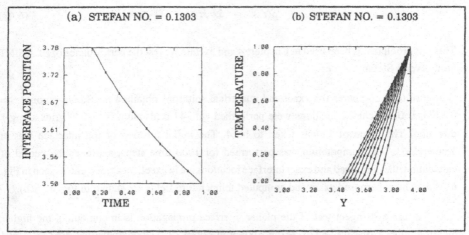

Figure 6.28 For Stefan No. = 0.1303, comparison of exact and computed values. (a) Interface position against time. (b) Temperature along Y at different times

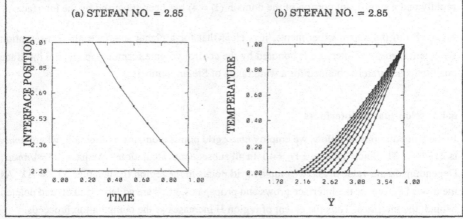

Figure 6.29 For Stefan No. =2.85, comparison of exact and computed values. (a) Interface position against time. (b) Temperature along Y at different times.

6.4.4   Zero Surface Tension

The behavior of interfaces with zero surface tension has been under scrutiny, with calculations indicating cusp–formation or tip–splitting (Bensimon and Pelce 1986). Since, in the absence of surface tension, no stabilization mechanism exists at any length scale, i.e., there is no smoothing effect, the disturbances of all wavelengths are amplified. Thus finite–time singularities can form

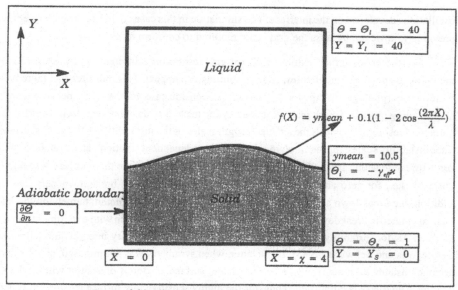

Figure 6.30. Illustration of computational domain and boundary conditions in nondimensional form.

on the interface. The evolution of an isothermal interface was tracked on grids with $nmax$=21,41 and 81 and the results for the finest grid is shown in Fig. 6.31. Clearly, for the isothermal interface the results are governed to a large degree by the grid spacing. Our calculations show that on the $nmax$=21 grid, the perturbation grows for the period of calculation and loses symmetry. The source of asymmetry comes from the unidirectional procedure used to define the interface shape and temperature gradients. For $nmax$=41, the interface first loses its symmetry, similar to the 21 grid case, and as time elapses the perturbation wavelengths permitted by the grid lead to a tip–splitting instability. A trough is formed upon tip–splitting, and rapid accumulation of latent heat ensues, leading quickly to the formation of a cusp in that region. Thus the sharp corner created there cannot be smoothed out in the absence of surface tension. For the finest grid, $nmax$=81, in Fig. 6.31, the interface becomes unstable to grid–scale oscillations. These short wavelength oscillations grow faster than the disturbances that are captured on the coarser grid and a cusp rapidly forms at the boundary where the imposed periodicity condition becomes incompatible with the asymmetric breakdown of the interface. The results on this fine grid correspond to our previous results obtained by employing boundary–fitted adaptive grids. There, in the absence of surface tension, a cusp formed at the boundary, and short wavelength oscillations developed on the interface. The simulation in this case could be carried farther than previously, however, because the grid is not required to conform to the interface.

It is interesting that with the different grid resolutions, different modes of instability including asymmetry, singularities, and tip–splitting can appear. This indicates that there is no preferred morphological shape for the zero surface tension case. Furthermore, because there is no prevailing physical length scale contained by the instability development itself, in effect the numerical resolution controls the smallest length scales of the instability. As the grid is refined, finer instability scales appear, implying that no grid independent solution can be obtained with zero surface tension. It was found in most of our calculations where the interface was highly unstable (i.e., for zero and low surface tensions) that asymmetric breakdown persisted. In addition, the breakdown always occurred on the same side of the domain, namely on the left. Such asymmetric breakdowns have been remarked upon in connection with experiments on the Saffman-Taylor instability (Tabeling et al. 1987) and in the boundary integral simulations of DeGregoria and Schwartz (1985). In the latter, when symmetry was not imposed, the interface became unstable asymmetrically. When sufficient surface tension is present, it will be shown that the asymmetry is eliminated. Thus, for the highly sensitive low surface tension cases, any small noise generated in the course of the computation leads to a breakdown of the smooth perturbation.

6.4.5 Low Surface Tension

The sensitivity of the solution procedure can be demonstrated by adding a surface tension γ_{eff} $=10^{-3}$ where, as defined previously, $\gamma_{eff} = \varepsilon_3 \Theta_m \bar{\gamma}$, Eq. (4.40). Since the curvature is $O(1)$, the

Gibbs-Thomson condition will lead to interface temperature variations of $O(10^{-3})$. This modification of the temperature at the interface is extremely small when compared to the nondimensionalized undercooling value $\Theta_l = -40$ imposed at the liquid boundary. When this small value of surface tension is added, the $nmax=81$ calculation does not lead to a cusp at the boundary, and the interface remains smooth. By comparing Figs. 6.31 and 6.32, it is seen that even such small γ_{eff} eliminates the singularities at the interface. However, the perturbations along the interface continue to grow.

Developing further on the issue of surface tension as a subtle mechanism, we impose, continuing from the final stage shown in Fig. 6.31, a surface tension $\gamma_{eff} = 10^{-2}$. The result is displayed in Fig. 6.33. The interface perturbations have been completely suppressed by surface tension. Also, interestingly, the resulting stable finger growth appears to be regaining symmetry, as can be seen from the plots of velocity and interfacial derivatives, which are smooth in this case but for residual short wavelength modes from the instabilities in the formative period. The

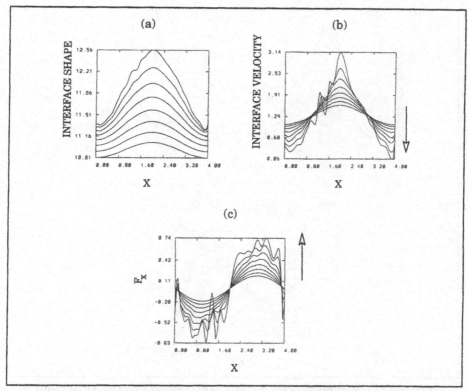

*Figure 6.31. Development of the zero surface tension interface on an nmax=81 grid.
The computations were stopped at t=0.8 due to cusp formation at the boundary.
(a) Interface shape plotted at equal time intervals ($\Delta t=0.1$). (b) Velocity along the
interface plotted at the same time instants. (c) f_x along the interface against x. Arrows
show sequence in time.*

different results demonstrate that the numerical technique maintains the history of the interface
with satisfactory accuracy.

For low surface tension cases, computational studies that impose symmetry may result
in interface patterns different from those without such a constraint, due to the highly
path-dependent morphology of the interface. Any change in the boundary conditions or
computational details will create different numerical noises, which in turn yield different final
interface shapes. To illustrate the impact of enforcing symmetry on the interface for low surface
tension, we compute the development of only half the perturbation by imposing a symmetry
condition for the other half. For the *nmax*=81 case, with low surface tension $\gamma_{eff} =10^{-3}$, the

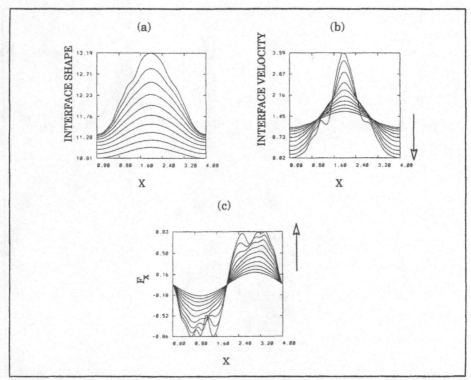

Figure 6.32. Development of the interface on the nmax=81 grid for surface tension $\gamma_{eff}=0.001$ up to time t=0.8. No cusp is formed. (a) Interface shape plotted at equal time intervals ($\Delta t=0.1$). (b) Velocity along interface at same times. (c) f_x along the interface against x.

resulting symmetric structure is shown in Fig. 6.34. There is again a series of instabilities in the vicinity of the tip, reminiscent of the structures observed in the Saffman-Taylor experiments with tip bubbles (Couder et al. 1986). On a finer grid, *nmax*=161, a similar phenomenon results, as shown in Fig. 6.35, except that in this case the tip breaks down at shorter wavelengths and the instability development is much more rapid than in the case of the coarser grid calculations. Thus, imposition of symmetry does not influence the stability of this low surface tension interface, however, the shape of the computed interface is very different from the case where symmetry is not explicitly imposed. Nevertheless, the wavelengths in both Fig. 6.34 and Fig. 6.35 are quite comparable, which is consistent with the fact that grid resolution is the main cut-off length scale with zero or low surface tension. It should be clarified, however, that as long

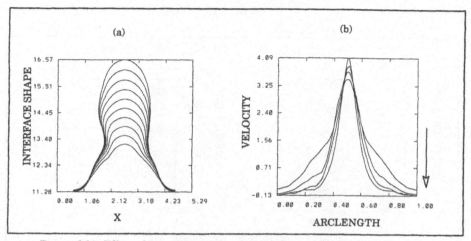

Figure 6.33. Effect of increasing surface tension for nmax=81 grid. Surface tension γ_{eff} = 10^{-2}. Interface evolution from t=0.8 to t=2.0. The initial condition is the final stage in Fig. 6.31 (for γ_{eff} = 10^{-3}, t=0.8). (a) Interface shapes at equal intervals of time showing stabilization and symmetrization. (b) Velocity along the interface at four different time instants showing symmetrization.

as surface tension is non-zero, there exists a physical length scale to control the morphological length scale of the interface. The results discussed so far, however, are for very low surface tensions, which creates a length scale too small to be resolved by the grid spacing employed here. When such low values of the surface tension are used, one has to be careful in interpreting results from a coarse grid calculation, since the physical features obtained are dependant on grid size.

6.4.6 Stable Fingers for Significant Surface Tension

The extreme sensitivity of the low surface tension cases to grid spacing, which yield widely different behaviors upon refinement, prompts us to confirm grid independence for the more stable higher surface tension cases. As already explained, this exercise can be conducted only for non–negligible surface tension cases.

Figure 6.36 compares the interface shape and velocity of an interface with γ_{eff} = 10^{-2}, *nmax*=41 (full lines), and *nmax*=81 (open circles). While in the initial stages of development, the profiles as well as interface velocities are in agreement; as the perturbation develops to large amplitudes, the coarser grid calculation underestimates the instability magnitude. In fact, the final stage of the fine grid calculation shows a vastly different value of velocity. Figure 6.37

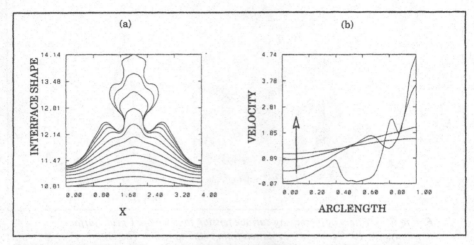

Figure 6.34. Effect of imposition of symmetry on development of interface for low surface tension ($\gamma_{eff} = 10^{-3}$) on nmax=81 grid. (a) Interface shapes shown after reflection. (b) Velocity along left half of interface against normalized arclength.

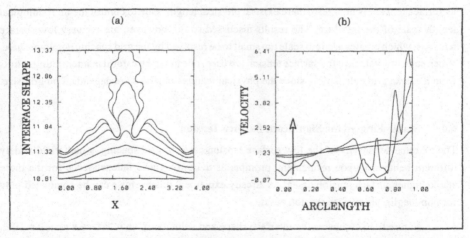

Figure 6.35. Effect of imposition of symmetry on development of interface for low surface tension ($\gamma_{eff} = 10^{-3}$) on nmax=161 grid. (a) Interface shown after reflection. (b) Velocity along left half of interface against normalized arclength.

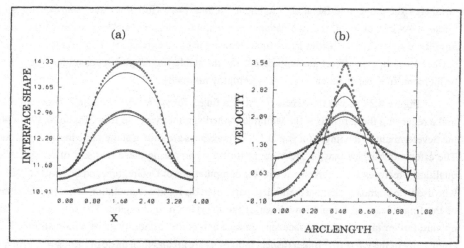

Figure 6.36. Comparison of solutions for surface tension $\gamma_{eff} = 10^{-2}$ on nmax=41 (full lines) and nmax=81 (open circles) grids. (a) Interface shapes at different time instants. (b) Interface velocities at same times as in (a).

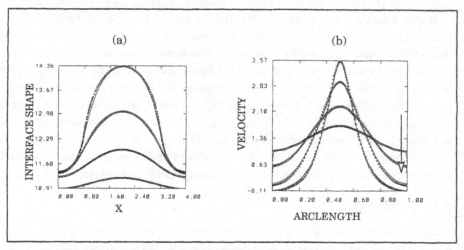

Figure 6.37. Comparison of solutions for surface tension $\gamma_{eff} = 10^{-2}$ on nmax=81 (full lines) and nmax=161 (open circles) grids. (a) Interface shapes at different time instants. (b) Interface velocities at same times as in (a).

compares the same interfacial development for *nmax*=81 (full line) and *nmax*=121 (open circles). As can be seen, the two calculations maintain close consistency, even in the large amplitude stage. The velocities in the final stage are in close agreement. Thus, for high enough surface tension, grid effects are suppressed, and the stable finger growth converges under grid refinement. In subsequent calculations we employ *nmax*=81.

Figure 6.38 shows the development of a finger for $\gamma_{eff} = 10^{-2}$. The first observation to make regarding these results is the well maintained symmetry of the front. Figure 6.39 shows the development of the finger for γ_{eff}=0.1. The interfaces are shown at the same instants of time. The effects of surface tension are brought forth by comparing these two sets of results. The qualitative features are in agreement with the computations of DeGregoria and Schwartz (1986), who used the boundary element method. In particular, the rapid accumulation of heat on the sides of the finger leads to a rapid slowdown of the interface in that region. The front propagates upward farther in the higher surface tension case before the instability gathers momentum. The amplitudes of the $\gamma_{eff} = 10^{-2}$ finger (aspect ratio $\approx$ 4) are greater than those of the $\gamma_{eff} = 0.1$ finger (aspect ratio $\approx$ 2.5), demonstrating the higher degree of instability for lower surface tension. As can be seen in the plot of the interfacial curvatures against x (Fig. 6.39 (c)), the higher surface tension (10^{-1}) causes the finger to spread laterally, leading to a multiple-valued interface with respect to x. The sides of the lower surface tension (10^{-2}) finger are almost flat and vertical, indicating minimal lateral spreading. The circular arc fits for representing the interfacial segments hold up very well, even for such a flat vertical surface. As can be seen by comparing the plots of curvature against x, the finger in the higher surface tension case has a more rounded tip and there is a wide region near the tip where the curvature is nearly a constant. In contrast, for the case of γ_{eff}=10^{-2}, the finger is sharper at the tip. The interfaces in both cases appear to have attained stable, shape-preserving growth. This may also be seen for the γ_{eff}= 0.1 case from the plots of tip curvature with time, shown in Fig. 6.39 (d). However, as observed by Saffman and Taylor (1958), over a substantial length and time the growth velocity of the finger is not constant and the tip is still accelerating, due to the finite domain size in our calculation. As the finger reaches an asymptotically invariant shape, the tip velocity will approach constant values.

6.5 SUMMARY

As demonstrated in this chapter, it is now possible to track unstable, multiple-valued interfaces with conduction heat transfer in the solid and liquid phases. We approached the simulation by first developing an explicit interface tracking method employing markers with connecting circular arc linkages. Employing circular arcs offered certain advantages in obtaining such interface features as curvature, slopes, and direction of normals. The interface was tracked on a Cartesian grid. In order to assess the efficacy of the method, several tests were designed.

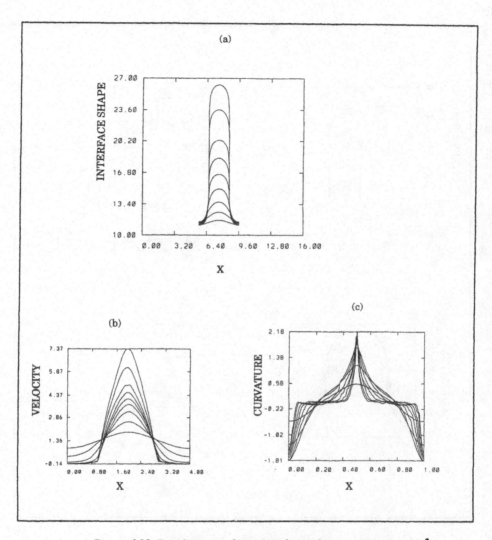

Figure 6.38. Development of interface for surface tension $\gamma_{eff} = 10^{-2}$ up to t=4.0. (a) Interface shapes at equal intervals of time. The finger has been drawn to scale. (b) Velocity along interface against x. (c) curvature along interface against x.

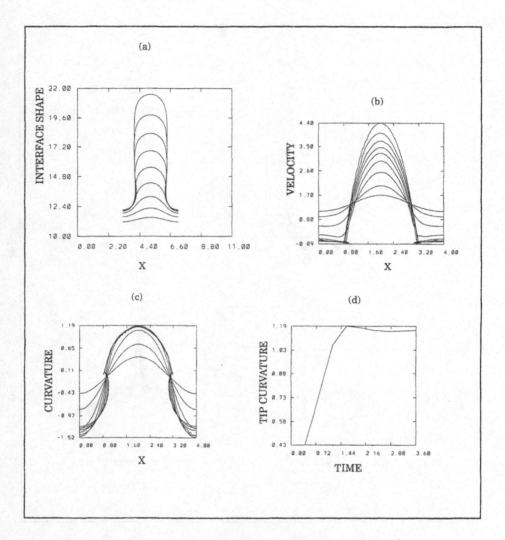

Figure 6.39. Development of interface for surface tension $\gamma_{eff} = 10^{-1}$ up to t=4.0. (a) Interface shapes at equal intervals of time. The finger has been drawn to scale. (b) Velocity along interface against x. (c) Curvature along interface against x. (d) Variation of tip curvature with time.

Several of the limitations usually associated with explicit interface tracking schemes have been resolved. The use of cells has facilitated the tracking of an interface that increases in perimeter with time. New points are automatically incorporated on the interface via an intersection procedure, thus circumventing the problems and corrective measures accompanying marker particle depletion/accumulation. It is envisaged that the intersection procedure will aid in solving the field equations also. A strategy has been developed, applying fairly simple, one-dimensional data arrays to simulate mergers and breakups of interfaces. This step has removed the most compelling limitation attached to Lagrangian tracking schemes. With the inclusion of the Eulerian grid feature and some strategies to lighten the logical burden, performing the merger/breakup tasks has been facilitated. The test cases tried thus far have successfully executed the expected merger and breakup tasks.

The next step in our simulation efforts involved integrating the new approach to interface tracking with the field equation solver. The field solver needs to operate on a fixed Cartesian grid layout and transfer information regarding front velocity to the interface tracking module. The front velocity is obtained from the Stefan condition upon solution of the field equations. In turn, the interface shape and position serve to impose internal boundary conditions for the field solver.

We developed the control volume formulation for tracking highly distorted fronts on fixed Cartesian grids under the influence of transport phenomena described by the Navier-Stokes equations. The facility afforded by the Cartesian grid in terms of setting up the control volume formulation leads to a conservative heat flux treatment across the moving interface. The execution of these procedures is fairly simple and involves dealing with a one-dimensional array of interfacial cells. Non-interfacial cells are assembled as usual. Thus, in contrast with other Eulerian methods, it is possible here to explicitly specify the location and shape of the interface and to apply the boundary conditions at the exact location of the interface. In strictly Lagrangian methods, on the other hand, the grid translates with the interface and needs to be periodically smoothed and redistributed. The non-boundary-fitted grid layout here circumvents such problems. Several issues have been resolved concerning the way in which the fluxes are computed at the moving interface, and the control volumes are assembled to enforce conservation of fluxes between adjoining cells. This is done by defining partner cells. Application of boundary conditions at the interface is achieved within the framework of the SIMPLE algorithm, by modifying the coefficients that arise in setting up the discretized form of the governing equations. However, unlike in the purely Eulerian methods, these boundary conditions are not lumped into source terms.

One difficulty for numerical solution methods for moving boundary problems is the paucity of analytical or experimental results for verifying the accuracy of the numerics. The highly non-linear, coupled nature of the moving boundary problems have rendered exact solutions very difficult to obtain. However, for the pure conduction situation, the method has been tested first against available exact solutions. For a planar interface, the Neumann solutions for a melting problem have been accurately reproduced by the numerical solutions with both moving and fixed grid methods, the latter including both enthalpy formulation and ELAFINT. The results for the deformed interface, at the morphological scale, illustrate clearly the effects of surface tension. In the absence of surface tension, singularities quickly develop on the front for sufficient grid resolution. The interaction of grid resolution, which governs the wavelength of noise on the interface with the surface propagation, is seen to affect the interfacial evolution. In particular, at very low surface tension the disturbances allowed by the grid strongly influence the interface motion and morphology. Thus, care needs to be exercised in interpreting results in this range of surface tension parameter. For sufficient interfacial tension, the results have been demonstrated to converge under grid refinement. The initial perturbation develops in time into long fingers, as in the Saffman-Taylor problem. An explicit front tracking method, such as ELAFINT, is needed for such problems. The qualitative features of the solutions obtained from ELAFINT are in agreement with other simulations of the fingering phenomenon. The finger shapes reach steady-state, while the tip continues to accelerate due to the fixed domain size and boundary conditions. In the next chapter, comparison is made between ELAFINT and the body-fitted/enthalpy methods for problems including convective transport in the liquid phase.

CHAPTER 7

ASSESSMENT OF FIXED GRID TECHNIQUES

7.1   INTRODUCTION

In this chapter, we compare the solutions to the Navier-Stokes equations obtained for stationary as well as moving boundary problems. The methods considered are the curvilinear coordinate-based method, with the enthalpy formulation if phase change is present, and ELAFINT which employs Cartesian grids and explicit interface tracking. The formulations for these methods are provided in the previous chapters. The scope of the tests conducted here is not restricted to moving boundary problems. Flow configurations involving stationary, irregular geometries also are computed. It is demonstrated that both curvilinear, body-fitted grids and the ELAFINT algorithm based on Cartesian coordinates can be effectively used to handle irregularly shaped, external as well as internal boundaries.

7.2   RESULTS FOR STATIONARY BOUNDARIES

We first compare solutions for a stationary interface with those from a procedure using body-fitted coordinates (Shyy 1994, Shyy et al. 1985) reviewed in Chapter 2, and the ELAFINT algorithm discussed in Chapter 6. This exercise is designed to validate the conservation and consistency characteristics of the cut cell method. The interface is sufficiently deformed that all the types of cut cells are encountered. In the absence of a consistent discretization in the vicinity of the interface, it is found that convergence cannot be achieved, and that care needs to be exercised in performing flux calculations in the grid cells affected by the interface. The computational domain is as shown in Fig. 7.1(a). Also shown in Fig. 7.1(c) and (d) are the schematics of the grids used for the computations using the fixed grid and boundary-fitted grid methods. The Navier-Stokes equations given in Chapter 2, Eqs. (2.1a–d), are solved. The square cavity is a frequently adopted test bed for numerical experiments for incompressible flows and benchmarks exist. We deform the base of the cavity, the amplitude being 10 percent of the base. A 121 x 121 Cartesian grid is employed. We first present the results for a driven cavity flow, where the top wall of the cavity is pulled at velocity $U = 1$ corresponding to a Reynolds number of 1000. The results from the present method are compared with the body-fitted formulation discussed in Chapter 2, with the same grid size. The results are shown in Fig. 7.2. The streamline patterns in Fig. 7.2(a) and (b) show good agreement. A quantitative comparison can be seen in Fig. 7.2(c) and (d), where the centerline velocities are plotted.

In Fig. 7.3, we compare the results for a stationary interface with natural convection in the cavity shown in Fig. 7.1(b). The governing equations are given in Chapter 2, Eqs. (2.1a–d), with the Boussinesq approximation for the buoyancy term. The same grid size and interface shape as above is used. The Rayleigh number computed is 10^5 and Pr = 1. Again the streamline patterns shown in Fig. 7.3(a) and (b) are in good agreement. In Fig. 7.3(e), (f), (g) we compare the centerline values of u, v components of velocity and temperature, respectively, along the centerline of the cavity. The results agree closely with the body-fitted code. Thus, it has been shown that the current scheme yields accurate results for the case of a stationary interface, in the presence of complex flowfields. It is noted that the formulation developed would be useful in computing incompressible flows around complex shapes employing fixed Cartesian grids, circumventing the need for generation of boundary-fitted grids.

7.3 MELTING FROM A VERTICAL WALL

We now proceed to test the numerical procedure developed here for a situation involving phase change. Hitherto, much effort has been devoted to numerically duplicating the results of Gau and Viskanta (1986) (hereafter referred to as G&V) for the melting of gallium from a vertical wall in a rectangular enclosure. Gallium is adopted as the experimental material since it is a metal with low melting point and thus is easy to handle. Unfortunately, numerical values of interface positions were not presented in G&V, and the initial conditions are ambiguous. The authors also present interface shapes viewed from the front and rear of their experimental set-up, and poor correspondence is observed. Furthermore, recent experiments of Campbell et al. (1994) appear to differ from G&V in regard to interface shapes and positions. Underlying these facts is the difficulty in performing experiments in relation to flowfields and interface positions in opaque melts. Despite these limitations, the experiments of G&V have been extensively employed for comparison. Needless to say, the agreement between the numerics and experiment is at best

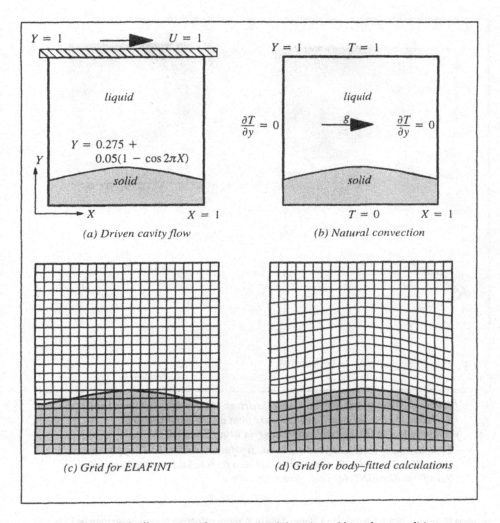

$Y = 1$ $\longrightarrow$ $U = 1$

liquid

$Y = 0.275 +$
    $0.05(1 - \cos 2\pi X)$

Y

solid

X $X = 1$

(a) Driven cavity flow

$Y = 1$ $T = 1$

liquid

$\dfrac{\partial T}{\partial y} = 0$ $\xrightarrow{\;g\;}$ $\dfrac{\partial T}{\partial y} = 0$

solid

$T = 0$ $X = 1$

(b) Natural convection

(c) Grid for ELAFINT

(d) Grid for body–fitted calculations

Figure 7.1. *Illustration of computational domain and boundary conditions for the test cases presented. For both configurations, no-slip conditions are applied on all solid surfaces. A Cartesian grid is used, with the interface passing through the grid. The interface amplitude was equal to 0.1, which is 10% of the domain size. (a) Domain and boundary conditions for driven cavity flow. (b) For natural convection. (c)Schematic of the grid arrangement for the ELAFINT computations. (d) Schematic of the grid for the body-fitted computations.*

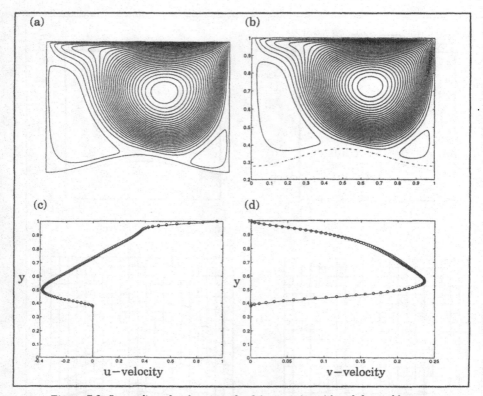

Figure 7.2. Streamlines for the case of a driven cavity with a deformed base. Re=1000, 121 × 121 grid. (a) Contours from a boundary-fitted grid computation. (b) On a fixed Cartesian grid using ELAFINT.
Comparison of solutions from the boundary–fitted and ELAFINT approaches. (c) Centerline u-velocity. Dots represent solution from boundary-fitted grid. Full lines from ELAFINT. (d) centerline v-velocity.

modest (Lacroix 1989, Lacroix and Voller 1990). A more effective comparison may be between numerical techniques of essentially disparate nature, for instance, purely Lagrangian and Eulerian methods. Lacroix and Voller (1990) have performed such a comparison. The grid sizes used by these authors, however, may not be sufficiently fine to resolve all the flow features. In our work, we found the presence of multiple convection cells in the initial stages of development of the interface in some cases. It is not conceivable that such cells can be resolved by coarse grids. The presence of such cells is important to capture, in particular because the interface shape reflects the presence of these flow features. Thus, two different numerical schemes can yield

252

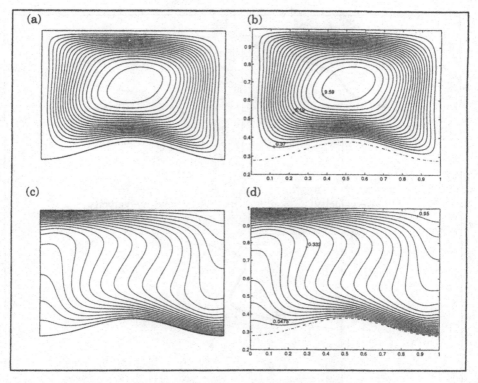

Figure 7.3. Streamlines for natural convection in a deformed cavity. $Re=10^5$, $Pr=1$, 121×121 grid. (a) Result from boundary-fitted grid computations. (b) Results from ELAFINT. (c) Temperature contours from the boundary-fitted formulation. (d) Temperature contours from ELAFINT.

results in good agreement for the same grid spacing, but neither may actually be an accurate calculation. In fact, the level of numerical dissipation, i.e., the order of accuracy of the numerical technique, was found to determine the types of flow features resolvable, especially at higher Rayleigh numbers. Thus, the entire situation, especially for the higher Rayleigh numbers, is found to be highly sensitive to the numerical conditions employed. Since the interface shape is strongly linked to the flow features and vice versa, great care must be employed in performing the computations.

Here, we present calculations for the melting of gallium from a vertical wall for the configuration shown in Figure 7.4. The current method is compared with an enthalpy-based Eulerian method (Shyy and Rao 1994a) employing an 81 x 81 grid. Unless otherwise mentioned,

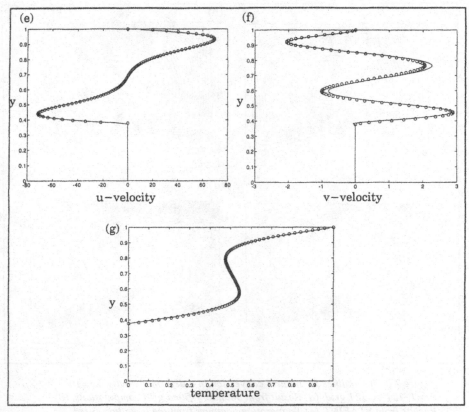

Figure 7.3 — continued. (e) Centerline u-velocity. Dots are from the boundary-fitted grid computations. Full lines are from ELAFINT. (f) Centerline v-velocity comparison. (g) Centerline temperatures.

the second-order central difference scheme is used for both the methods. In a method dealing with the temperature as a variable and using an interface tracking procedure, it is not possible to initiate melting in a domain that is entirely solid. Thus, the computations using the present method are started from an initial condition generated by the enthalpy method so that a thin, initial melt layer exists at the start. The flowfield and temperature field are obtained from the purely Eulerian method. Melting is initiated at the left wall. In Fig. 7.5, we show the results for a Rayleigh number of 10^4. The Prandtl number of gallium is 0.021. The Stefan number for this case is 0.042. The interfacial shape and position are compared in this case with that of the enthalpy-based method in Fig. 7.5(a). As can be seen, close agreement is maintained between

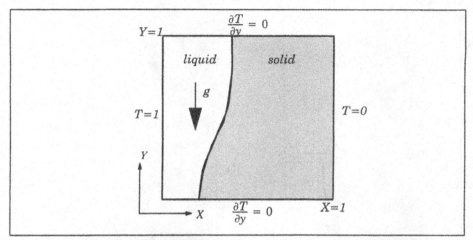

Figure 7.4. Illustration of computational domain and boundary conditions for the melting of gallium. The melting is initiated at the left wall. No-slip velocity conditions are imposed on all solid surfaces. Computations are performed on a Cartesian grid.

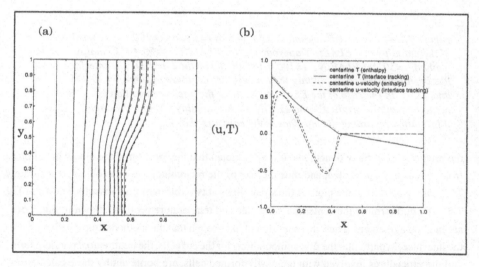

Figure 7.5. Melting of gallium, Ra=10⁴, Pr=0.021, St=0.042, 81x81 grid. Melting is initiated at the left. (a) Comparison of interface positions at equal intervals of time. Dotted lines represent the enthalpy-based method; full lines correspond to ELAFINT. (b) Comparison of centerline profiles. Legends are self-explanatory.

255

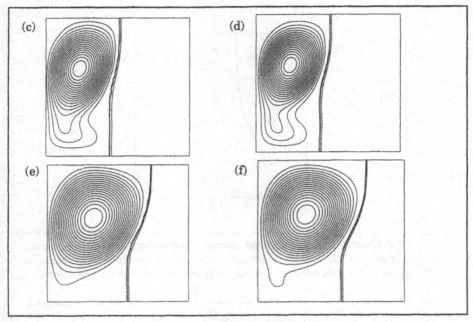

Figure 7.5 — continued. Streamline contours at t=50 and t=100. Comparison of the enthalpy and ELAFINT approaches. (c) ELAFINT, t=50. (d) Enthalpy method, t=50. (e) ELAFINT, t=100. (f) Enthalpy method, t=100. The interface has been represented by plotting the temperature contours between T=−0.005 and 0.005. In the case of the ELAFINT method, the interface position is actually available explicitly and exactly. In the enthalpy method, the only information regarding the interface is the contours shown.

the predictions of the two methods. It may be noted that the location of the elbow in the phase front is correctly predicted, and thus the size of the recirculation zone is obtained accurately. This can be seen from the plots of the streamlines at two different time instants, shown in Fig. 7.5(c–f). Furthermore the centerline velocities and temperature profiles, shown in Fig. 7.5(b), are in good agreement. Thus, this experiment has proven that the methods employed in treating the interface, in particular the flux computations for the cut cells, the treatment of pressure terms, and the procedures involved with the newly formed cells, are borne out by these calculations.

We next present (Fig. 7.6) the results for a higher Stefan number. In this case, as above, Ra= 10^4, Pr=0.021, but St=0.42, a ten-fold increase. In Fig. 7.6(a), the interface positions are compared and plotted at equal intervals of time $\delta\tau = 10$. The results are very close, except for the lagging of the current method for larger times. But even for the rapidly moving interface,

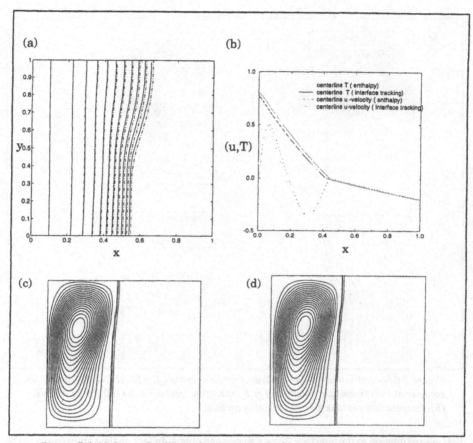

Figure 7.6. Melting of gallium. Comparison of solutions from an enthalpy method and ELAFINT methods. Higher Stefan number case: $Re=10^4$, $Pr=0.021$, $St=0.42$, 81x81 grid. (a) Comparison of interface shapes at equal intervals of time. Full lines correspond to ELAFINT. Dashes represent enthalpy method. (b) Comparison of centerline values. The legend is self-explanatory. (c) Streamline contours for ELAFINT at $t=50$. (d) Streamline contours for enthalpy method at $t=100$. In (c) and (d), the interface is represented by plotting the temperature contours $T=-0.005$, 0.0, and 0.005.

the front shapes are well predicted, which implies that the bulk flow, including recirculation zones, is obtained accurately. In the figures containing contours of streamfunction and isotherm contours, Fig. 7.6(c–h), the interface has been represented in each case by plotting the temperature contours $\Theta = -0.005$, 0.0, and 0.005. In the case of the enthalpy method, this is the

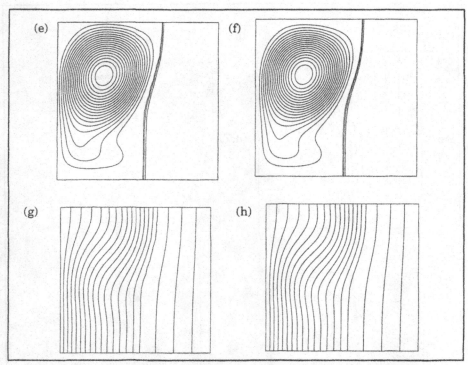

Figure 7.6 — continued. (e) Streamline contour at t=100, for ELAFINT. (f) Streamline contour at t=100, enthalpy method. (g) Temperature contour at t=100 for ELAFINT. (h) Temperature contour, t=100, enthalpy method.

best approximation to the interface shape that one can obtain. In fact, in the comparisons of the interface shapes, the values corresponding to the enthalpy method are obtained as that corresponding to the $f = 0.5$ contour by interpolation. Thus there is an uncertainty of the order of the grid spacing in identifying the interface position in the enthalpy method. In contrast, the interface tracking procedure explicitly yields information regarding interface shape. Also, it is noted that in the interface tracking method, no artificial modeling of physics is necessary in the vicinity of the interface, whereas in the enthalpy method there is an unavoidable smearing of information. In addition, the D'arcy law treatment and the mushy zone model (Voller and Prakash 1987) obscure the transport processes in this region.

The results presented here demonstrate the accuracy of the computational procedure by comparing with previously tested methods. Thus, at this stage we have a computational procedure that can handle the evolution of highly distorted fronts. At the microscale level, at which diffusion processes dominate, the method can handle complex geometries and physics. At the macroscopic level, the flow features obtained are in agreement with the solutions from body-fitted techniques for deformed boundaries. Further extension of this work is required to combine all the capabilities demonstrated here to incorporate the entire range of phenomena encountered in the solidification process.

With the addition of convection, the solutions obtained from ELAFINT have been compared to the boundary-fitted approach with the enthalpy method. The results have been demonstrated to be in close agreement. It is clear that the different types of algorithms presented in this work have their individual characteristics in terms of accuracy and efficiency. It is hoped that the methodologies and examples presented in this work will aid in providing a better understanding of the issues involved in simulating incompressible flows with moving boundaries.

7.4   SUMMARY

Purely Lagrangian and Eulerian methods for interface tracking have their characteristic advantages and disadvantages. A combination of these two approaches provides a flexible and general computational capability for solving moving boundary problems. With this idea in mind, we have developed and assessed several ideas and resolved various issues associated with tracking an evolving, moving front over an underlying Eulerian grid. The phenomenon of solidification is chosen to illustrate the procedure for the following reasons:

1.   The discontinuous moving boundary separates two phases with distinctly different physical behavior.

2.   The moving boundary is a source of latent heat.

3.   Curvature dependant boundary conditions apply on the interface.

4.   The interface is driven by the gradients of the transported variable.

5.   In the case of unstable solidification, the interface becomes extremely distorted, sometimes leading to changes in topology.

6.   The phenomenon has been extensively investigated and many of the physical issues resolved to a fair degree.

7.   It is both a significant technological and theoretical issue.

It has been demonstrated that the ELAFINT technique, by drawing upon the strengths of the purely Eulerian as well as Lagrangian aspects, provides a powerful new tool for the investigation of a variety of complex flow phenomena. In particular, its ability to tackle the phase change problem in the pure material case, where there is a phase discontinuity, is noteworthy.

7.5 CONCLUDING REMARKS

A variety of numerical methods have been presented in this book to solve moving boundary problems in fluid mechanics. The interaction of the flow with moving phase fronts and with flexible/elastic structures such as membranes has been discussed as applications of these methods. Both macroscopic and microscopic aspects have been covered when dealing with phase change problems, and the distinct computational requirements in each case have been discussed. Both the curvilinear coordinate and ELAFINT approaches can be treated as general purpose solvers for fluid flows in complex geometries. The several numerical examples and test problems presented in the previous chapters support the applicability of these two techniques in fluid flows containing a sharp interface or a curved moving boundary. With the addition of other techniques such as multigrid methods and adaptive/composite grid procedures, both the accuracy and efficiency of these general fluid flow solvers can be enhanced. Work on these aspects is in progress, and it is envisaged that the effectiveness and versatility of the techniques introduced in this work will be further advanced by the incorporation of these enhancements.

REFERENCES

Aidun, C.K. 1991, Principles of hydrodynamic instability: application in coating systems, 3 Parts, *TAPPI J.*, Vol. 74, pp. 209–220.

Ananth, R. and Gill, W. N. 1989, Dendritic growth of an elliptic paraboloid with forced convection in the melt, *J. Fluid Mech.*, Vol. 208, pp. 575–593.

Ananth, R. and Gill, W. N. 1991, Self-consistent theory of dendritic growth with convection, *J. Crystal Growth*, Vol. 108, pp. 173–189.

Anderson, D.A.,Tannehill, J.C. and Pletcher, R.H. 1984, *Computational Fluid Mechanics and Heat Transfer*, Hemisphere, Washington, DC.

Ashgriz, N. and Poo, J. Y. 1991, FLAIR: Flux Line-Segment Model for Advection and Interface Reconstruction, *J. Comp. Phys.*, Vol. 93, pp. 449–468.

Baker, A.J. 1983, *Finite Element Computational Fluid Mechanics*, Hemisphere Publication Co., Washington, DC.

Batina, J.T. 1989, Unsteady Euler airfoil solutions using unstructured dynamic meshes, *AIAA Paper No. 89–0115*.

Bear, J. 1988, *Dynamics of Fluids in Porous Media*, Dover, New York.

Beckermann, C. and Viskanta, R. 1993, Mathematical modeling of transport phenomena during alloy solidification, *Appl. Mech. Rev.*, Vol. 46, pp. 1–27.

Beckermann, C., Feller, R. J., Irwin, T.R., Muller-Spath, H. and Wang, C.Y. 1994, Visualization of sedimentation and thermosolutal convection during equiaxed alloy solidification, *AIAA Paper No. 94–0570*.

Bejan, A., 1984, *Convection Heat Transfer*, Wiley, New York.

Bejan, A., 1994, Contact melting heat transfer and lubrication, in J.P. Hartnett, T.F. Irvine, Jr. and Y.I. Cho (eds.), *Advances in Heat Transfer*, Vol. 24, pp. 1–38, Academic Press, New York.

Bell, J.B., Collela, P. and Glaz, H.M. 1989, A second–order prediction method for the incompressible Navier–Stokes equations, *J. Comp. Phys.*, Vol. 85, pp. 257–283.

Ben-Jacob, E., Godbey, R., Goldenfeld, N. D., Koplik, J., Levine, H., Mueller, T. and Sander, L. M. 1985, Experimental demonstration of the role of anisotropy in interfacial pattern formation, *Phys. Rev. Lett.*, Vol. 55, pp. 1315–1318.

Bennon, W. D. and Incropera, F. P. 1987, A continuum model for momentum, heat and species transport in binary solid-liquid phase change systems – I. Model formulation, *Intl. J. Heat Mass Transf.*, Vol. 30, pp 2161–2170.

Bensimon D. 1986, Stability of viscous fingering, *Phys. Rev. A*, Vol. 33, pp. 1302–1308.

Bensimon, D. and Pelce, P. 1986, Tip-splitting solutions to the Stefan problem, *Phys. Rev. A*, Vol. 33, pp. 4477–4478.

Bentley, W.A. and Humphreys, W.J. 1962, *Snow Crystals,* Dover, New York (originally published by McGraw–Hill, New York, 1931).

Bisplinghoff, R., Ashley, H. and Halfman, R. 1955, *Aeroelasticity,* Addison–Wesley, Reading, MA.

Bogy, D.B. 1979, Drop formation in a circular liquid jet, *Ann. Rev. Fluid Mech.*, Vol. 11, pp. 207–228.

Borisov, A. G., Fedorov, O. P., and Maslov, V. V. 1991, Growth of Succinonitrile dendrites in different crystallographic directions, *J. Crystal Growth*, Vol. 112, pp. 463–466.

Boschitsch, A. H. and Quackenbush, T R. 1993, High accuracy computation of fluid–structure interaction in transonic cascades, *AIAA Paper No. 93–0485*, presented at 31st Aerospace Sciences Meeting.

Bouissou, Ph., Perrin, B. and Tabeling, P. 1990, Influence of an external periodic flow on dendritic crystal growth, in F. H. Busse and L. Kramer (eds.), *Nonlinear Evolution of Spatio-temporal Structures in Dissipative Dynamical Systems,* Plenum Press, New York.

Braaten, M. and Shyy, W. 1986a, Comparison of iterative and direct solution methods for viscous flow calculations in body–fitted coordinates, *Int. J. Numer. Meths. Fluids*, Vol. 6, pp. 325–349.

Braaten, M. and Shyy, W. 1986b, A study of recirculating flow computation using body-fitted coordinates: Consistency aspects and mesh skewness, *Num. Heat Transf.*, Vol. 9, pp. 559–574.

Brackbill, J.U., Kothe, D.B. and Zemach, C. 1992, A continuum method for modeling surface tension, *J. Comp. Phys.*, Vol. 100, pp. 335– 354.

Brattkus, K. 1989, Capillary instabilities in deep cells during directional solidification, *J. Phys. France*, Vol. 50, pp. 2999–3006.

Brebbia, C.A., Telles, J.C.F. and Wrobel, L.C. 1984, *Boundary Element Techniques*, Springer–Verlag, New York.

Brenner, E. A. 1991, Pattern selection in two-dimensional dendritic growth, *Advances in Phys.*, Vol. 40, pp. 53–97.

Brice, J.C. 1986, *Crystal Growth Processes*, Blackie, London, U.K.

Brower, R., Kessler, D., Koplik, J. and Levine, H. 1983, Geometric approach to moving-interface dynamics, *Phys. Rev. Lett.*, Vol. 51, pp. 1111–1114.

Brown, R. A. 1988, Theory of transport processes in single crystal growth from the melt, *AIChE J.*, Vol. 34, pp. 881–911.

Brush, L. N. and Sekerka, R.F. 1989, A numerical study of a two-dimensional crystal growth forms in the presence of anisotropic growth kinetics, *J. Crystal Growth*, Vol. 96, pp. 419–441.

Cahn, J. W. and Hilliard, J. E. 1958, Free energy of a nonuniform system, I. Interfacial free energy, *J. Chem. Phys.*, Vol. 28, pp. 258–267.

Campbell, T. A., Pool, R. E. and Koster, J. N. 1994, Melting and solidification of a liquid metal at a vertical wall, *AIAA Paper No. 94–0792*.

Canuto, C., Hussaini, M.Y., Quarteroni, A. and Zang, T.A. 1988, *Spectral Methods in Fluid Dynamics*, Springer-Verlag, New York.

Carey, G.F. and Oden, J.T. 1986, *Finite Elements (Volume 6): Fluid Mechanics*, Prentice–Hall, Englewood–Cliffs, NJ.

Carey, V. P. 1992, *Liquid-Vapor Phase-Change Phenomena*, Hemisphere Publishing Co., Washington, DC.

Carpenter, B. M. and Homsy, G.M. 1989, Combined buoyant-thermocapillary flow in a cavity, *J. Fluid Mech.*, Vol. 207, pp. 121–132.

Chambers, L. I. 1966, A variational formulation of the Thwaites sail equation, *Quarterly J. Mech. Appl. Math.*, Vol.19, pp. 221–231.

Chan, C.L. and Mazumder, J. 1987, One–dimensional steady–state model for damage by vaporization and liquid expulsion due to laser–material interaction, *J. Appl. Phys.*, Vol. 62, pp. 4579–4585.

Chan, R. K.–C. and Street, R.L. 1970, A computer study of finite amplitude water waves, *J. Comp. Phys.*, Vol. 62, pp. 4579–4585.

Cheer, A.Y. and van Dam, C.P. (eds.) 1993, *Fluid Dynamics in Biology*, American Mathematical Society, Providence, R.I.

Chen, C.P., Shang, H.M. and Jiang, Y. 1992, An efficient pressure-velocity procedure for gas-droplet two-phase flow calculations, *Int. J. Numer. Meths. Fluids*, Vol. 15, pp. 233–245.

Chen, G. and Zhou, J. 1992, *Boundary Element Methods*, Academic Press, New York.

Chen, H., Saghir, M.Z. and Chehab, S. 1994, Numerical study on transient convection in float zone induced by g–jitters, *J. Crystal Growth*, Vol. 142, pp. 362–372.

Chen, M.M., 1987, Thermocapillary convection in materials processing, in Samanta, S.K., Komanduri, R., McMeeking, R., Chen, M.M. and Tseng, A. (eds.) 1987, *Interdisciplinary Issues in Materials Processing and Manufacturing*, Vol. 2, pp. 541–558, ASME, New York.

Chen, S., Johnson, D.B. and Raad, P.E. 1995 Velocity boundary conditions for the simulation of free surface fluid flow, *J. Comp. Phys.*, Vol. 116, pp. 262–276.

Chiang, K.C. and Tsai, H.L. 1992a, Shrinkage-induced fluid flow and domain change in two-dimensional alloy solidification, *Int. J. Heat Mass Transf.*, Vol. 35, pp. 1763–1770.

Chiang, K.C. and Tsai, H.L. 1992b, Interaction between shrinkage-induced fluid flow and natural convection during alloy solidification, *Int. J. Heat Mass Transf.*, Vol. 35, pp. 1771–1778.

Chorin, A.J. 1967, A numerical method for solving incompressible viscous flow problems, *J. Comp. Phys.*, Vol. 2, pp. 12–26.

Chorin, A.J. 1968, Numerical solution of the Navier-Stokes equations, *Math. Comp.*, Vol. 22, pp. 745–762.

Christoudoulou, K.N. and Scriven, L.E. 1989, The fluid mechanics of slide coating, *J. Fluid Mech.*, Vol. 208, pp. 321–354.

Ciarlet, P.G. and Sanchez-Palencia, E. 1987, *Applications of Multiple Scaling in Mechanics*, Masson, Paris, France.

Cladis, P. E., Finn, P. L. and Gleeson, J. T. 1990, Routes to cell formation and hidden ramps in directional solidification, in *Nonlinear Evolution of Spatio-Temporal Structures in Dissipative Continuous Systems*, Ed. Busse, F. H. and L. Kramer (eds.), Plenum Press, New York.

Clift, R., Grace, J.R. and Weber, M.E. 1978, *Bubbles, Drops, and Particles*, Academic Press, New York.

Concus, P. 1990, Capillary surfaces in microgravity, in J. N. Koster and R. L. Sani (eds.), *Low-Gravity Fluid Dynamics and Transport Phenomena*, Vol. 130, Progress in Astronautics and Aeronautics, AIAA, Washington DC., pp. 183–206.

Coriell, S.R., Hardy, S.C. and Cordes, M.R. 1977, Stability of liquid zones, *J. Colloid Interface Sci.*, Vol. 60, pp. 126 – 136.

Couder, Y., Gerard, N. and Rabaud, M. 1986, Narrow fingers in the Saffman-Taylor instability, *Phys. Rev. A*, Vol. 34, pp. 5175–5178.

Crank, J. 1984, *Free and Moving Boundary Problems*, Oxford University Press, Oxford, U.K.

Cross, M.C. and Hohenberg, P.C. 1993, Pattern formation outside of equilibrium, *Rev. Mod. Phys.*, Vol. 65, pp. 851–1112.

Daly, B.J. 1967, Numerical study of two-fluid Rayleigh Taylor instability, *Phys. Fluids*, Vol. 10, pp. 297–307.

Daly, B.J. 1969, A technique for including surface tension in hydrodynamic calculations, *J. Comp. Phys*, Vol.4, pp. 97–117.

Darolia, R. 1991, NiAl alloys for high–temperature structural applications, *JOM*, March Issue, pp. 44–49.

Davis, S.H. 1990, Hydrodynamic interactions in directional solidification, *J. Fluid Mech.*, Vol. 212, pp. 241–262.

Davis, S.H., Muller, U. and Dietsche, C. 1984, Pattern selection in single-component systems coupling Benard convection and solidification, *J. Fluid. Mech.*, Vol. 144, pp. 133–151.

de Cheveigne, S., Faivre, G., Guthmann, C. and Kurowski, P. 1990, Directional solidification of transparent eutectic alloys, in F.H. Busse and L. Kramer (eds.), *Nonlinear Evolution of Spatiotemporal Structures in Dissipative Continuous Systems*, Plenum Press, New York.

DeGregoria, A.J. and Schwartz, L. W. 1985, Finger breakup in Hele-Shaw cells, *Phys. Fluids*, Vol. 28, pp. 2313–2314.

DeGregoria, A. J. and Schwartz, L. W. 1986, A boundary integral method for two-phase displacement in Hele-Shaw cells, *J. Fluid Mech.*, Vol. 164, pp. 383–400.

de Groh, H.C., III and Yao, M. 1994, Numerical and experimental study of transport phenomena in directional solidification of succinonitrile, *Transport Phenomena in Solidification*, C. Beckermann, H.P. Wang, L.A. Bertram, M.S. Sohal and S.I. Guceri (eds.), HTD–Vol. 284 & AMD–Vol. 182, ASME, New York, pp. 227–243.

de Matteis, G. and de Socoi, L. 1986, Nonlinear aerodynamics of a two-dimensional membrane airfoil with separation, *AIAA J.*, Vol. 23, pp. 831–836.

Demirdzic, I. and Peric, M. 1990, Finite volume method for prediction of fluid flow in arbitrary shaped domains with moving boundaries, *Int. J. Numer. Meths. Fluids*, Vol. 10, pp. 771–790.

Diao, Q.Z. and Tsai, H.L. 1993, The formation of negative and positive segregated bands during solidification of aluminum-copper alloys, *Int. J. Heat Mass Transf.*, Vol. 36, pp. 4299–4305.

Ding, Z. and Anghaie, S. 1994, Modelling of R–12 bulk evaporation and condensation in an encapsulated container, ASME 1994 Intl. Mechanical Engineering Congress and Exposition, Chicago, IL, Paper No. 94/WA HT–12.

Dowell, E.H. 1980, *A Modern Course in Aeroelasticity*, Sijthoff and Noordhoff, Amsterdam, The Netherlands.

Drazin, P.G. and Reed, H. W. 1981, *Hydrodynamic Stability*, Cambridge University Press, Cambridge, U.K.

Dukowicz, J.K. and Dvinsky, A.S. 1992, Approximate factorization as a high order splitting for the implicit incompressible flow equations, *I. Comp. Phys.*, Vol. 102, pp. 336–347.

Edwards, J.W., Bennett, R.M., Whitlow, W., Jr. and Seidel, D.A. 1983, Time–marching transonic flutter solutions including angle–of–attack effects, *J. Aircraft*, Vol. 20, pp. 899–906.

Emerton, N. E. 1984, *The Scientific Reinterpretation of Form*, Cornell University Press, Ithaca, NY.

Evans, G. and Greif, R. 1994, A two–dimensional model of the chemical vapor deposition of silicon nitride in a low–pressure hot–wall reactor including multicomponent diffusion, *Int. J. Heat Mass Transf.*, Vol. 37, pp. 1535–1544.

Fabietti, L.M., Seetharaman, V. and Trivedi, R. 1990, The development of solidification microstructures in the presence of lateral constraints, *Metall. Trans. A*, Vol. 21A, pp. 1299–1310.

Fang, R.T. , Glicksman, M.E., Coriell, S.R., McFadden, G.B. and Boisvert, R.F. 1985, Convective influence on the stability of a cylindrical solid-liquid interface, *J. Fluid Mech.*, Vol. 202, pp. 339–366, 1985.

Fauci, L.J. and Peskin, C. S. 1988, A computational model of acquatic animal locomotion, *J. Comp. Phys.*, Vol. 77, pp. 85–108.

Favier, J.J. 1990, Recent advances in Bridgeman growth modeling and fluid flow, *J. Crystal Growth*, Vol. 99, pp. 18–29.

Favier, J.J. and Rouzaud, A. 1983, Morphological stability of the solidification interface under convective conditions, *J. Crystal Growth*, Vol. 64, pp. 367–379.

Felicelli, S.D., Heinrich, J.C. and Poirier, D.R. 1991, Simulation of freckles during vertical solidification of binary alloys, *Metall. Trans. B*, Vol. 22B, pp. 847–859.

Filipovic, J., Viskanta, R. and Incropera, F.P. 1994, An analysis of subcooled turbulent film boiling on a moving isothermal surface, *Int. J. Heat Mass Transf.*, Vol. 37, pp. 2661–2674.

Finn, R. 1986, *Equilibrium Capillary Surfaces*, Springer–Verlag, New York.

Flemings, M. C. 1974, *Solidification Processing*, McGraw–Hill, New York.

Flesselles, J.-M., Simon, A.J. and Libchaber, A.J. 1991, Dynamics of one-dimensional interfaces: an experimentalist's view, *Adv. Phys.*, Vol. 40, pp. 1–51.

Fletcher, C.A.J. 1988, *Computational Techniques for Fluid Dynamics*, Springer–Verlag, New York, 2 volumes.

Floryan, J. M. and Rasmussen, H. 1989, Numerical methods for viscous flows with moving boundaries, *Appl. Mech. Rev.,* Vol. 42, pp. 323–341.

Forth, S. A. and Wheeler, A. A. 1989, Hydrodynamic and morphological stability of the unidirectional solidification of a freezing binary alloy: A simple model, *J. Fluid Mech.,* Vol. 202, pp. 339–366.

Fromm, J. 1981, Finite difference computation of the capillary jet, free surface problem, in *Lecture Notes in Physics,* Vol. 238, Springer Verlag, New York, pp. 188–193, 1981.

Fukai, J., Zhao, Z., Poulikakos, D., Megaridis, C.M. and Miyatake, O. 1993, Modeling of the deformation of a liquid droplet impinging upon a flat surface, *Phys. Fluids,* Vol. A5, pp. 2588–2599.

Fukai, J., Shiiba, Y., Yamamoto, T., Miyatake, O., Poulikakos, D., Megaridis, C.M. and Zhao, Z. 1995, Wetting effects on the spreading of a liquid droplet colliding with a flat surface: experiment and modeling, *Phys. Fluids,* Vol. 7, pp. 236–247.

Gau, G. and Viskanta, R. 1986, Melting and solidification of a pure metal on a vertical wall, *J. Heat Transf.,* Vol. 108, pp. 174–181.

Geyling, F.T. and Homsy, G.M. 1990, Extensional instabilities of the glass fibre drawing process, *Glass Tech.,* Vol. 21, pp. 95–102.

Glicksman, M.E., Coriell, S. R. and McFadden, G. B. 1986, Interaction of flows with the crystal-melt interface, *Ann. Rev. Fluid Mech.,* Vol. 18, pp. 307–336.

Glimm, J., Grove, J., Lindquist, B., McBryan, O. A. and Tryggvason, G. 1988, The bifurcation of tracked scalar waves, *SIAM J. Sci. Stat. Comput.,* Vol. 9, pp. 61–79.

Glimm, J., McBryan, O., Melnikoff, R. and Sharp, D.H. 1986, Front tracking applied to Rayleigh-Taylor instability, *SIAM J. Sci. Stat. Comput.,* Vol. 7, pp. 230–251.

Gottlieb, D. and Orszag, S.A. 1977, *Numerical Analysis of Spectral Methods,* SIAM, Philadelphia, PA.

Greenhalgh S., Curtiss, H.C. and Smith, B. 1984, Aerodynamic properties of two dimensional inextensible flexible airfoils, *AIAA J.,* Vol. 22, pp. 865–870.

Halpern, D. and Grotberg, J.B. 1992, Fluid–elastic instabilities of liquid–lined flexible tubes, *J. Fluid Mech.,* Vol. 244, pp. 615–632.

Harlow, F.H. and Welch, J.E. 1965, Numerical calculation of time–dependent viscous incompressible flow of fluid with free surface, *Phys. Fluids,* Vol. 8, pp. 2182–2189.

Heinrich, J.C., Felicelli, S. and Poirier, D.R. 1991, Vertical solidification of dendritic binary alloys, *Comp. Meths. Appl. Mech. Engrg.,* Vol. 89, pp. 435–461.

Hirata, T., Makino, Y. and Kaneko, Y. 1991, Analysis of close–contact melting for octadecane and ice inside isothermally heated horizontal rectangular capsule, *Int. J. Heat Mass Transf.,* Vol. 34, pp. 3097–3106.

Hirsch, C. 1988, 1990, *Numerical Computation of Internal and External Flows,* Wiley, New York, 2 volumes.

Hirt, C.W. and Nichols, B.D. 1981, Volume of Fluid (VOF) method for the dynamics of free boundaries, *J. Comp. Phys,* Vol. 39, pp. 201–225.

Hirt, C.W., Amsden, A.A. and Cook, J.L. 1974, An arbitrary Lagrangian–Eulerian computing method for all speeds, *J. Comp. Phys,* Vol. 14, pp. 227–253.

Homsy, G.M. 1987, Viscous fingering in porous media, *Ann. Rev. Fluid Mech.*, Vol. 19, pp. 271–312.

Hosangadi, A., Merkle, C.L. and Turns, S.R., 1990, Analysis of Forced Combusting Jets, *AIAA J.*, Vol. 28, pp. 1473–1480.

Hou, T.Y., Lowengrub, J.S. and Shelley, M.J. 1994, Removing the stiffness from interfacial flows with surface tension,, *J. Comp. Phys.*, Vol. 114, pp. 312–338.

Howison, S.D., Ockendon, J.R. and Lacey, A.A. 1985, Singularity development in moving-boundary problems, *Quart. J. Mech. Appl. Math.*, Vol. 38, pp. 343–360.

Huang, S.-C. and Glicksman, M.E. 1981, Fundamentals of dendritic solidification – I. Steady state tip growth, *Acta Metall.*, Vol. 29, pp. 701 – 715. Fundamentals of dendritic solidification – II. Development of side branch structure, *Acta Metall.*, Vol. 29, pp. 717–734.

Huerre, P. 1987, Spatio-temporal instabilities in closed and open shear flows, in E. Tirapegui and D. Villaroel (eds.), *Instabilities and Nonequilibrium Structures*, pp. 141–177, Dodrecht, Reidel, Germany.

Huerta, A. and Liu, W.K. 1988, Viscous flow with large free surface motion, *Comput. Meths. Appl. Mech. Engrg.*, Vol. 69, pp. 277–324.

Hughes, T.J.R. 1987, *The Finite Element Method*, Prentice–Hall, Englewood–Cliffs, NJ.

Hughes, T.J.R., Liu, W.K. and Brooks, A. 1979, Finite element analysis of incompressible viscous flow by the penalty formulation, *J. Comp. Phys.*, Vol. 30, pp. 1–60.

Hughes, T.J.R. and Hulbert, G.M. 1988, Space–time finite element methods for elastodynamics: formulations and error estimates, *Comput. Meths. Appl. Mech. Engrg.*, Vol. 66, pp. 339–363.

Hurle, D.T.J. (ed.) 1993, *Handbook of Crystal Growth*, Vol. 1 A and B, North–Holland, Amsterdam, the Netherlands.

Issa, R.I. 1985, Solution of the implicit discretized fluid flow equations by operator-splitting, *J. Comp. Phys.*, Vol. 62, pp. 40–65.

Jackson, P. S. 1983, A Simple model for elastic two-dimensional elastic sails, *AIAA J.*, Vol. 21, pp. 153–155.

Jaluria, Y. and Torrance, K.E. 1986, *Computational Heat Transfer*, Hemisphere Publishing Co., Washington, DC.

Kamotani, Y., Ostrach, S. and Pline, A. 1994, Analysis of velocity data in surface tension driven convection experiment in microgravity, *Phys. Fluids*, Vol. 6, pp. 3601–3609.

Kanouff, M. and Greif, R. 1994, Oscillations in thermocapillary convection in a square cavity, *Int. J. Heat Mass Transf.*, Vol. 37, pp. 885–892.

Kar, A. and Mazumder, J. 1990, Two–dimensional model for material damage due to melting and vaporization during laser irradiation, *J. Appl. Phys.*, Vol. 68, pp. 3884–3891.

Kar, A., Rockstroh, T. and Mazumder, J. 1992, Two–dimensional model for laser–induced materials damage: effects of assist gas and multiple reflections inside cavity, *J. Appl. Phys.*, Vol. 71, pp. 2560–2569.

Kassemi, M. and Naraghi, M.H.N. 1994, Combined heat transfer and fluid flow analysis of semi–transparent crystals in low-g and 1-g solidification, *Transport Phenomena in*

Solidification, C. Beckermann, H.P. Wang, L.A. Bertram, M.S. Sohal and S.I. Guceri (eds.), HTD–Vol. 284 & AMD–Vol. 182, ASME, New York, pp. 245–254.

Kaviany, M. 1991, *Principles of Heat Transfer in Porous Media,* Springer-Verlag, New York.

Kaviany, M. 1994, *Principles of Convective Heat Transfer,* Springer-Verlag, New York.

Keene, B. J. 1993, Review of data for the surface tension of pure metals, *Intl. Metals Reviews,* Vol. 38, pp. 157–192.

Kessler, D.A., Koplik, J. and Levine, H. 1988, Pattern selection in fingered growth phenomena, *Advances in Physics,* Vol. 37, pp. 255–339.

Klausner, J.F., Mei, R., Bernhard, D.M. and Zeng, L.Z. 1993, Vapor bubble departure in forced convection boiling, *Int. J. Heat Mass Transf.,* Vol. 36, pp. 651–662.

Koai, K., Damaschek, R. and Bergmann, H.W. 1993, Heat transfer in laser hardening of rotating cylinders, in C.L. Chan, F.P. Incropera and V. Prasad (eds.), *Transport Phenomena in Nonconventional Manufacturing and Materials Processing,* HTD–Vol. 259, ASME, New York, pp. 1–8.

Kobayashi, R 1993, Modelling and numerical simulations of dendritic crystal growth, *Physica D,* Vol. 63, pp. 410–423.

Kopf-Sill, A.R. and Homsy, G.M. 1988, Bubble motion in a Hele–Shaw cell, *Phys. Fluids,* Vol. 31, pp. 18–26.

Koschmieder, E.L. 1993, *Benard Cells and Taylor Vortices,* Cambridge University Press, Cambridge, U.K.

Kothe, D.B. and Mjolsness, R.C. 1992, RIPPLE : A new method for incompressible flows with free surfaces, *AIAA J.,* Vol. 30, pp. 2694–2700.

Kothe, D.B. and Rider, W.J. 1994, Comments on modeling interfacial flows with volume-of-fluid methods, *Los Alamos National Laboratory Report No. LA–UR–94–3384,* Los Alamos, NM.

Kuo, T.-W. 1991, Three–dimensional computations of flow and fuel injection in an engine intake port, ASME Paper No. 91–ICE–4.

Kurowski, P., de Cheveigne, S., Faivre, G. and Guthman, C. 1989, Cusp instability in cellular growth, *J. Phys. France,* Vol. 50, pp. 3007–3019.

Kurz, W. and Fischer, D. J. 1989, *Fundamentals of Solidification,* Trans. Tech. Publications, Aerdermannsdorf, Switzerland.

Kurz, W. and Trivedi, R. 1992, Microstructure and phase selection in laser treatment of materials, *Trans. ASME,* Vol. 114, pp. 450–458.

Kwak, D., Chang, J.L.C., Shanks, S.P. and Chakravarthy, S.R. 1986, A three-dimensional incompressible Navier-Stokes flow solver using primitive variables, *AIAA J.,* Vol. 24, pp. 390–396.

Lacroix, M. 1989, Computation of heat transfer during melting of a pure substance from an isothermal wall, *Numer. Heat Transf.,* Part B, Vol. 15, pp. 191–210.

Lacroix, M. and Voller, V. R. 1990, Finite difference solutions of solidification phase change problems: transformed versus fixed grids, *Numer. Heat Transf., Part B,* Vol. 17, pp. 25–41.

Lafaurie, B., Nardone, C., Scardovelli, R., Zaleski, S. and Zanetti, G. 1994, Modelling merging and fragmentation in multiphase flows with SURFER, *J. Comp. Phys.*, Vol. 113, pp. 134–147.

Lai, Y.G. and Przekwas, A.J. 1994, A finite-volume method for fluid flow simulations with moving boundaries, *Comp. Fluid Dyn.*, Vol. 2, pp. 19–40.

Lamb, H. 1932, *Hydrodynamics*, 6th ed., Cambridge University Press, Cambridge, U.K. (also 1945, Dover, New York.)

Lan, C.W. and Kou, S. 1991a, Thermocapillary flow and natural convection in a melt column with an unknown melt/solid interface, *Int. J. Numer. Meths. Fluid*, Vol. 12, pp. 59–80.

Lan, C.W. and Kou, S. 1991b, Heat transfer, fluid flow and interface shapes in floating–zone crystal growth, *J. Crystal Growth*, Vol. 108, pp. 351–366.

Langbein, D. 1990, Crystal growth from liquid column, *J. Crystal Growth*, Vol. 104, pp. 47–59.

Langbein, D. 1992, Stability of liquid bridges between parallel plates, *Microgravity Sci. and Tech.*, Vol. 5, pp. 2–11.

Langer, J.S. 1980, Instabilities and pattern formation in crystal growth, *Rev. Mod. Phys.*, Vol. 52, pp. 1–56.

Langer, J.S. and Muller-Krumbhaar, H. 1978, Theory of dendritic growth, Parts I, II, and III, *Acta Metall.*, Vol. 28, pp. 1681–1708.

Langlois, W.E. 1985, Buoyancy-driven flows in crystal-growth melts, *Ann. Rev. Fluid Mech.*, Vol. 17, pp. 191–215.

Lefebvre, A. H. 1989, *Atomization and Sprays*, Hemisphere Publishing Co., Washington, DC.

Lee, D. and Chiu, J.J. 1992, Computation of physiological bifurcation flows using a patched grid, *Comp. Fluids*, Vol. 21, pp. 519–535.

Lee, R.C. and Nydahl, J.E. 1989, Numerical calculation of bubble growth in nucleate boiling from inception through departure, *J. Heat Transf.*, Vol. 111, pp. 474–479.

Lee, S. H.–K. and Jaluria, Y. 1993, Radiative transport in the cylindrical furnace for optical fiber drawing, in *Transport Phenomena in Nonconventional Manufacturing and Materials Processing*, C.L. Chan, F.P. Incropera and V. Prasad (eds.), HTD–Vol. 259, ASME, New York, pp. 43–58.

Levich, V.G. 1962, *Physicochemical Hydrodynamics*, Prentice–Hall, Englewood Cliffs, NJ.

Liang, P. Y. 1991, Numerical method for calculation of surface tension flows on arbitrary grids, *AIAA J.*, Vol. 29, pp. 161–167.

Li, J., Sun, J. and Saghir, Z. 1993, Buoyant and thermocapillary flow in liquid encapsulated floating zone, *J. Crystal Growth*, Vol. 131, pp. 83–96.

Li, J. and Saghir, M.Z. 1994, Coupled buoyant and surface tension driven convection in liquid encapsulation floating zone under 1-g and 0-g with deformable interfaces, AIAA Paper No. AIAA–94–0793.

Löhner, R., Yang, C., Cebral, J., Baum, J.D., Luo, H., Pelessone, D. and Charman, C. 1995, Fluid–Structure Interaction Using a Loose Coupling Algorithm and Adaptive Unstructured Grids, AIAA Paper No. 95–2259.

Lu, P.-J., Pan, D. and Yeh, D.-Y. 1995, Transonic flutter suppression using active acoustic excitation, *AIAA J.*, Vol. 33, pp. 694–702.

Lynch, D. R. 1982, Unified approach to simulation on deforming elements with application to phase change problem, *J. Comp. Phys.*, Vol. 47, pp. 387–411.

Marchaj, C.A. 1979, *Aero-Hydrodynamics of Sailing*, Dodd, Mead and Co., New York, NY.

Maxworthy, T.M. 1986, Bubble formation, motion and interaction in a Hele-Shaw cell, *J. Fluid Mech.*, Vol. 173, pp. 95–114.

Mazumder, J. 1991, Overview of melt dynamics in laser processing, *Optical Engrg.*, Vol. 30, pp. 1208–1219.

McCormick, B.W., 1979, *Aerodynamics, Aeronautics and Flight Mechanics*, Wiley, New York.

McFadden, G.B., Coriell, S.R., Boisvert, R.F., Glicksman, M.E. and Fang, Q.T. 1984, Morphological stability in the presence of fluid flow in the melt, *Metall. Trans. A*, Vol. 15A, pp. 2117–2124.

McLean, J.W. and Saffman, P.G. 1981, The effect of surface tension on the shape of fingers in a Hele–Shaw cell, *J. Fluid Mech.*, Vol. 102, pp. 455–469.

Mei, E., Chen, W. and Klausner, J.F. 1995 Vapor bubble growth in heterogeneous boiling – I. formulation; II. growth rate and thermal fields, *Int. J. Heat Mass Transf.*, Vol. 38, pp. 909–919 & pp. 921–934.

Meiburg, E. and Homsy, G.M. 1988, Nonlinear unstable viscous fingers in Hele–Shaw flows. II: numerical simulation, *Phys. Fluids*, Vol. 31, pp. 429–439.

Melaaen, M.C. 1992, Calculation of fluid flows with staggered and nonstaggered curvilinear nonorthogonal grids, *Numer. Heat Transf.*, Vol. 21, pp. 1–19.

Minkowycz, W.J., Sparrow, E.M., Schneider, G.E. and Pletcher, R.H. 1988, *Handbook of Numerical Heat Transfer,* Wiley, New York.

Miyata, H. 1986, Finite difference simulation of breaking waves, *J. Comp. Phys.*, Vol. 65, pp. 179–214.

Mizuta, Y. 1991, Generalized boundary conditions on the basis of a deformable–cell method: free surface, density interfaces and open boundaries, *Comp. Fluids*, Vol. 19, pp. 377–385.

Moallemi, M.K., Webb, B.W. and Viskanta, R. 1986, An experimental and analytical study of close–contact melting, *J. Heat Transf.*, Vol. 108, pp. 8940–899.

Mulder, W., Osher, S., and Sethian, J.A. 1992, Computing interface motion in compressible gas dynamics, *J. Comp. Phys.*, Vol. 100, pp. 209–224.

Mullins, W.W. and Sekerka, R.F. 1964, Stability of a planar interface during solidification of a dilute binary alloy, *J. Appl. Phys.*, Vol. 3, pp. 444–451.

Mundrane, M. and Zebib, A. 1994, Oscillatory buoyant thermocapillary flow, *Phys. Fluids*, Vol. 6, pp. 3294–3305.

Murai, H. and Maruyama, S. 1980, Theoretical investigation of the aerodynamics of double membrane sailwing airfoil sections, *J. Aircraft*, Vol. 17, pp. 294–329.

Murai, H. and Maruyama, S. 1982, Theoretical investigation of sailwing airfoils taking account of elasticities, *J. Aircraft*, Vol. 19, pp 385–389.

Murata, S. and Tanaka, S. 1989, Aerodynamic characteristics of a two–dimensional porous sail, *J. Fluid Mech.*, Vol. 206, pp. 463–475.

Murthy, J.Y. 1987, A numerical simulation of flow, heat and mass transfer in a floating zone at high rotational Reynolds numbers, *J. Crystal Growth*, Vol. 83, pp 23–34.

Myshkis, A.D., Babskii, V.G., Kopachevskii, N.D., Slobozhanin, L.A. and Tyuptsov, A.D. 1987, *Low Gravity Fluid Mechanics*, translated by R.S. Wadhwa, Springer–Verlag, New York.

Nadarajah, A. and Narayanan, R. 1986, A comparison result for convection in bounded geometries", *J. Math. Phys. (ZAMP)*, Vol. 37, pp. 280–283.

Nadarajah, A. and Narayanan, R. 1987, On the completeness of the Rayleigh–Marangoni and Graetz eigenspaces and the simplicity of their eigenvalues, *Quart. Appl. Math.*, Vol. XLV, pp. 81–92.

Nadarajah, A. and Narayanan, R. 1989, Morphological instability in dilute binary systems. I. A uniform approach, *Physico-Chemical Hydrody.*, Vol. II, pp. 81–102.

Nadarajah, A. and Narayanan, R. 1990, Comparison between morphological and Rayleigh–Marangoni instabilities, in D. Meinkoehn and H. Haken (eds.), *Dissipative Structures in Transport processes and Combustion*, Springer Verlag Series in Synergetics, Springer-Verlag, New York, Vol. 48, pp. 215–228.

Narayanan, R., Zhao, A.-X. and Wagner, C. 1992, Rayleigh–Marangoni bilayer convection in liquid encapsulated crystal growth, in S.H. Davis, G. Worster and H.E. Huppert (eds.), *Interactive Dynamics of Convection with Solidification*, NATO–ASI Series, Kluwer Publishers, Boston, MA, pp. 15–18.

Neilson, D.G. and Incropera, F.P. 1993, Effect of rotation on fluid motion and channel formation during unidirectional solidification of a binary alloy, *Int. J. Heat Mass Transf.*, Vol. 36, pp. 435–442.

Neilson, D.G. and Incropera, F.P. 1991, Unidirectional solidification of a binary alloy and the effects of induced fluid motion, *Int. J. Heat Mass Transf.*, Vol. 34, pp. 1717–1732.

Newman, B.G. 1987, Aerodynamic theory for membranes and sails, *Prog. Aero. Sci.*, Vol. 24, pp. 1–27.

Newman, B.G. and Low, H.T. 1984, Two-dimensional impervious sails: Experimental results compared with theory, *J. Fluid Mech.*, Vol. 144, pp. 445–462.

Ni, J. and Beckermann, C. 1993, Modeling of globulitic alloy solidification with convection, *J. Mater. Proc. Manuf. Sci.*, Vol. 2, pp. 217–231.

Nielsen, J. N. 1963, Theory of flexible aerodynamic surfaces, *J. Appl. Mech.*, Vol. 30, pp. 435–442.

Nobari, M.R.H. and Tryggvason, G. 1994a, Numerical simulations of drop collisions, *NASA Technical Memorandum*, 106751.

Nobari, M.R.H. and Tryggvason, G. 1994b, The flow induced by the coalescence of two initially stationary drops, *NASA Technical Memorandum*, 106752.

Nomura, T. 1981, ALE finite element computations of fluid-structure interaction problems, *Comput. Meths. Appl. Mech. Engrg.*, Vol. 29, pp. 329–349.

Oran, E.S. and Boris, J.P. 1987, *Numerical Simulation of Reactive Flow*, Elsevier, New York.

Orme, M. and Muntz, E.P. 1990, The manipulation of capillary stream breakup using amplitude–modulated disturbances: a pictorial and quantitative representation, *Phys. Fluids A*, Vol. 2, pp. 1124–1140.

Orszag, S.A. 1980, Spectral methods for problems in complex geometries, *J. Comp. Phys.*, Vol. 37, pp. 70–92.

Osher, S. and Sethian, J. A. 1988, Fronts propagating with curvature dependent speed: Algorithms based on Hamilton-Jacobi formulations, *J. Comp. Phys.*, Vol. 79, pp. 12–49.

Ostrach, S. 1983, Fluid mechanics in crystal growth, *J. Fluids Engrg.*, Vol. 105, pp. 5–20.

Ostrach, S., Kamotani, Y. and Lee, J. 1993, Oscillatory thermocapillary flows, *Adv. Space Res.*, Vol. 13, pp. 97–104.

Ouyang, H. and Shyy, W. 1995, Multi–zone simulation of Bridgman growth process of β-NiAl crystal, to be published.

Paek, U.C. and Runk, R.B. 1978, Physical behavior of the neck-down region during furnace drawing of silica fibers, *J. Appl. Phys.*, Vol. 49, pp. 4417–4422.

Park, C.-W., Maruvada, S.R.K. and Yoon, D.-Y. 1994, The influence of surfactant on the bubble motion in Hele-Shaw cells, *Phys. Fluids*, Vol. 6, pp. 3267–3275.

Patankar, S.V. 1980, *Numerical Heat Transfer and Fluid Flow*, Hemisphere Publishing Corp., Washington, DC.

Patankar, S.V. and Spalding, D.B. 1972, A calculation procedure for heat, mass and momentum transfer in three-dimensional parabolic flows, *Int. J. Heat and Mass Transf.*, Vol. 15, pp. 1787–1806.

Patera, A.T. 1984, A spectral element method for fluid dynamics: Laminar flow in a channel expansion, *J. Comp. Phys.*, Vol. 54, pp. 468–488.

Pelce, P. (ed.) 1988, *Dynamics of Curved Fronts*, in Perspectives in Physics Series, Academic Press, New York.

Penrose, O. and Fife, P.C. 1990, Thermodynamically consistent models of phase-field type for the kinetics of phase transitions, *Physica D*, Vol. 43, pp. 44–62.

Perry, R.H. (ed.) 1984, *Perry's Chemical Engineers' Handbook*, 6th ed., McGraw–Hill, New York.

Peyret, R. and Taylor, T.D. 1983, *Computational Methods for Fluid Flow*, Springer–Verlag, New York.

Pieters, R. and Langer, J. S. 1986, Noise-driven sidebranching in dendritic crystal growth, *Phys. Rev. Lett.*, Vol. 56, pp. 1948–1952.

Pinkus, O. 1990, *Thermal Aspects of Fluid Tribology*, ASME, New York, pp. 340–352.

Prakash, C. 1990, Two–phase model for binary solid-liquid phase change, part I: Governing equations, *Numer. Heat Transf.*, part B, Vol. 18, pp. 131–145.

Prakash, C. and Voller, V. 1989, On the numerical solution of continuum mixture model equations describing binary solid-liquid phase change, *Numer. Heat Transf.*, part B, Vol. 15, pp 171–189.

Preisser, F., Schwabe, D. and Scharmann, A. 1983, Steady and oscillatory thermocapillary convection in liquid columns with free cylindrical surface, *J. Fluid Mech.*, Vol. 126, pp. 545–567.

Prescott, P.J. and Incropera, F.P. 1993, Binary solid–liquid phase change with fluid flow, *Advances in Transport Processes*, A.S. Mujumdar and R.A. Mashelkar (eds.), Vol. IX, pp. 57–101.

Prescott, P.J. and Incropera, F.P. 1994, Convective transport phenomena during solidification of a binary metal alloy: I – Numerical predictions, *J. Heat Transf.*, Vol. 116, pp. 735–749.

Probstein, R.F. 1989, *Physicochemical Hydrodynamics*, Butterworths, Boston, MA. (2nd edition, 1994, Wiley, New York.)

Quon, D.H.H., Chehab, S., Aota, J., Kuriakose, A.K., Wang, S.S.B., Saghir, M.Z. and Chen, H.L. 1993, Float-zone crystal growth of Bismuth Germanate and numerical simulation, *J. Crystal Growth*, Vol. 134, pp. 266 – 274.

Rallison, J.M. 1984, The deformation of small viscous drops and bubbles in shear flows, *Ann. Rev. Fluid Mech.*, Vol.16, pp. 45–66.

Rallison, J.M. and Acrivos, A. 1978, A numerical study of the deformation and burst of a viscous drop in an extensional flow, *J. Fluid Mech.*, Vol. 89, pp. 191–200.

Ramaswamy, B. and Kawahara, M. 1987, Arbitrary Lagrangian–Eulerian finite element for unsteady, convective, incompressible viscous free surface fluid flow, *Int. J. Numer. Meths. Fluids*, Vol. 7, pp. 1053–1075.

Rappaz, M. 1989, Modelling of microstructure formation in solidification processes, *Int. Mat. Rev.*, Vol. 34, pp. 93–123.

Rappaz, M. and Thevoz, Ph. 1987, Solute diffusion model for equiaxed dendritic growth: Analytical solution, *Acta Metall.*, Vol. 35, pp. 2929–2933.

Rayleigh, Lord. 1899, *Scientific Papers*, Vol. 1, Cambridge University Press, Cambridge, U.K.

Rhie, C.M. and Chow, W.L. 1983, A numerical study of the turbulent flow past an isolated airfoil with trailing edge separation, *AIAA J.*, Vol. 21, pp. 1525–1532.

Riahi, D.N. and Walker, J.S., 1989, Float zone shape and stability with the electromagnetic body force due to a radio–frequency induction coil, *J. Crystal Growth*, Vol. 94, pp. 635–642.

Richards, J.R., Lenhoff, A.M. and Beris, A.N. 1994, Dynamic breakup of liquid–liquid jets, *Phys. Fluids*, Vol. 6, pp. 2640–2655.

Rizzetta, D.P. 1979, Time–dependent response of a two–dimensional airfoil in transonic flow, *AIAA J.*, Vol. 17, pp. 26–32.

Roache, P.J. 1972, *Computational Fluid Dynamics*, Hermosa Publishers, Albuquerque, NM.

Roosen, A.R. and Taylor, J.E. 1994, Modeling crystal growth in a diffusion field using fully faceted interfaces, *J. Comp. Phys.*, Vol. 114, pp. 113–128.

Rosen, M.J. 1989, *Surfactants and Interfacial Phenomena*, 2nd ed., Wiley, New York.

Rosner, D.E. 1986, *Transport Processes in Chemically Reacting Flow Systems*, Butterworths, Boston, MA.

Rubinstein, E.R. and Glicksman, M.E. 1990, Dendritic growth kinetics and structure, I. Pivalic acid, II. Camphene, *J. Crystal Growth*, Vol. 112, pp. 84–110.

Rubinstein, L.I. 1971, *The Stefan Problem*, American Mathematical Society, Providence, RI.

Ruschak, K.J. 1985, Coating flows, *Ann. Rev. Fluid Mech.*, Vol. 17, pp. 65–89.

Saffman, P.G. 1986, Viscous fingering in Hele-Shaw cells, *J. Fluid Mech.*, Vol. 173, pp. 73–94.

Saffman, P.G. and Taylor, G.I. 1958, The penetration of a fluid into a porous medium or Hele-Shaw cell containing a more viscous fluid, *Proc. Royal Soc. London*, A Vol. 245, pp. 312– 329.

Saito, Y., Goldbeck-Wood, G. and Muller-Krumbhaar, H. 1988, Numerical simulation of dendritic growth, *Phys. Rev. A*, Vol. 38, pp. 2148–2157.

273

Samanta, S.K., Komanduri, R., McMeeking, R., Chen, M.M. and Tseng, A. (eds.) 1987, *Interdisciplinary Issues in Materials Processing and Manufacturing*, 2 volumes, ASME, New York.

Sanz Andres, A. 1992, Static and dynamic response of liquid bridges, in *Microgravity Fluid Dynamics*, H.J. Rath (ed.), Springer-Verlag, Berlin, Germany, pp. 3–17.

Sato, T., Kurz, W. and Ikawa, K. 1987, Experiments on dendrite branch detachment in the Succinonitrile-Camphor alloy, *Trans. Japan Inst. Met.*, Vol. 28, pp. 1012–1021.

Schlichting, H. 1979, *Boundary Layer Theory*, McGraw–Hill, New York.

Sears, B., Narayanan, R., Anderson, T.J. and Fripp, A.L. 1992, Convection of Tin in a Bridgman system, I – flow characterization by effective diffusivity measurements, *J. Cryst. Growth.*, Vol. 125, pp. 404–414.

Sears, B., Anderson, T.J., Narayanan, R. and Fripp, A.L., 1993, Detection of solutal convection during diffusivity measurements of oxygen in liquid Tin, *Metall. Trans.*, Vol. 24B, pp. 91–100.

Sen, S. and Stefanescu, D.M. 1991, Melting and casting processes for high-temperature intermetallics, *JOM*, May Issue, pp. 30–32.

Sethian, J.A. 1990, Numerical algorithms for propagating interfaces: Hamilton-Jacobi equations and conservation laws, *J. Diff. Geom.*, Vol. 31, pp. 131–161.

Sethian, J.A. and Strain, J. 1992, Crystal growth and dendritic solidification, *J. Comp. Phys.*, Vol. 98, pp. 231–253.

Shamsundar, N. and Rooz, E. 1988, Numerical methods for moving boundary problems, in Minkowycz, W.J., Sparrow, E.M., Schneider, G.E. and Pletcher, R.H. (eds.), *Handbook of Numerical Heat Transfer*, Wiley, New York, pp. 747–786.

Shuen, J.-S., Chen, K.-H. and Choi, Y., 1993, A coupled implicit method for chemical non-equilibrium flows at all speeds, *J. Comp. Phys.*, Vol. 106, pp. 306–318.

Shyy, W. 1987, An adaptive grid method for Navier-Stokes flow computation, *Appl. Math. Comp.*, Vol. 21, pp. 201–219.

Shyy, W. 1994, *Computational Modelling for Fluid Flow and Interfacial Transport*, Elsevier, Amsterdam, The Netherlands.

Shyy, W. and Burke, J. 1994, A study of iterative characteristics of convective-diffusive and conjugate heat transfer problems, *Numer. Heat Transf.*, Vol. 26B, pp. 21–37.

Shyy, W. and Chen, M.-H. 1991a, Interaction of thermocapillary and natural convection flows during solidification: normal and reduced gravity conditions, *J. Crystal Growth*, Vol. 108, pp 247–261.

Shyy, W. and Chen, M.-H. 1991b, Double diffusive flow in enclosures, *Phys. Fluids A*, Vol. 3, pp 2592–2607.

Shyy, W. and Chen, M.-H. 1993, A study of buoyancy induced and thermocapillary flow of molten alloy, *Comput. Meths. Appl. Mech. Engrg.*, Vol. 105, pp 333–358.

Shyy, W., Liang, S.-J., and Wei, D.Y. 1994, Effect of dynamic perturbation and contact condition on edge-defined fiber growth characteristics, *Intl. J. Heat Mass Transf.*, Vol. 37, pp 977–987.

Shyy, W., Pang, Y., Hunter, G.B., Wei, D.Y. and Chen, M.-H. 1992b, Modelling of turbulent transport during continuous ingot casting, *Int. J. Heat Mass Transf.*, Vol. 35, pp. 1229–1245.

Shyy, W., Pang, Y., Hunter, G.B., Wei, D.Y. and Chen, M.-H. 1993c, Effect of turbulent heat transfer on continuous ingot solidification, *J. Eng. Mat. Tech.*, Vol. 115, pp. 8–16.

Shyy, W. and Rao, M.M. 1994a, Enthalpy based formulations for phase change problems with application to g-jitter, *AIAA Paper No. 93–2831, also Microgravity Sci. and Tech.*, Vol. 7, pp. 41–491.

Shyy, W. and Rao, M.M. 1994b, Calculation of meniscus shapes and transport processes in float zone, *Int. J. Heat Mass Transf.*, accepted for publication, (*Added in Proof*: 1995, Vol. 38, pp 2281–2295).

Shyy, W., Thakur, S. and Wright, J. 1992a, Second–order upwind and central difference schemes for recirculating flow computation, *AIAA J.*, Vol. 30, pp. 923–932.

Shyy, W., Tong, S.S. and Correa, S.M. 1985, Numerical recirculating flow calculation using a body fitted coordinate system, *Numer. Heat Transf.*, Vol. 8, pp 99–105.

Shyy, W., Udaykumar, H.S. and Liang, S.-J. 1993a, A study of meniscus formation with application to edge–defined fiber growth process, *Phys. Fluids. A*, Vol. 5, pp 2610–2623.

Shyy, W., Udaykumar, H.S. and Liang, S.-J. 1993b, An interface tracking method applied to morphological evolution during phase change, *Intl. J. Heat Mass Transf.*, Vol. 36, pp. 1833 – 1844.

Shyy, W. and Vu, T.C. 1991, On the adoption of velocity variable and grid system for fluid flow computation in curvilinear coordinates, *J. Comp. Phys.*, Vol. 92, pp. 82–105.

Siegel, R. and Howell, J.R. 1981, *Thermal Radiation Heat Transfer*, Hemisphere Publishing Co., Washington, DC.

Slattery, J.C. 1990, *Interfacial Transport Phenomena*, Springer–Verlag, New York.

Slobozhanian, L.A. and Perales, J.M. 1993, Stability of liquid bridges between equal disks in an axial gravity field, *Phys. Fluids A*, Vol. 5, pp. 1305–1314.

Smith, J.B. 1981, Shape instabilities in pattern formation in solidification: A new method for numerical solution of the moving boundary problem, *J. Comp. Phys.*, Vol. 39, pp. 112–127.

Smith, R.W. and Shyy, W. 1995a, A computational model of flexible membrane wings in steady laminar flow, accepted for publication in *AIAA J.*

Smith, R.W. and Shyy, W. 1995b, Computation of unsteady laminar flow over a flexible two-dimensional membrane wing, accepted for publication in *Phys. Fluids.*

Soulaimani, A., Fortin, M., Dhatt, G. and Ouellet, Y. 1991, Finite element simulation of two- and three-dimensional free surface flows, *Comput. Meths. Appl. Mech. Engrg.*, Vol. 86, pp. 265–296.

Sneyd, A.D. 1984, Aerodynamic coefficients and longitudinal stability of sail airfoils, *J. Fluid Mech.*, Vol. 149, pp. 127–146.

Spangler, C.A., Hilbing, J.H. and Heister, S.D. 1995, Nonlinear modeling of jet atomization in the wind-induced regime, *Phys. Fluids*, Vol. 7, pp. 964–971.

Sparrow, E.M. and Myrum, T.A. 1985, Inclination-induced direct-contact melting in a circular tube, *J. Heat Transf.*, Vol. 107, pp. 533–540.

Sparrow, E.M., Patankar, S.V. and Ramadhyani, S. 1977, Analysis of modeling in the presence of natural convection in the melt region, *J. Heat Transf.*, Vol. 99, pp. 520–526.

Stone, H.A. 1994, Dynamics of drop deformation and breakup in viscous fluids, *Ann. Rev. Fluid Mech.*, Vol. 26, pp. 65–102.

Stone, H.A. and Leal, L.G. 1989, Relaxation and breakup of an initially extended drop in an otherwise quiescent fluid, *J. Fluid Mech.*, Vol. 198, pp. 399–427.

Strain, J. 1989, A boundary integral approach to unstable solidification, *J. Comp. Phys.*, Vol. 85, pp. 342–389.

Sugimoto, T. and Sato, J. 1988, Aerodynamic characteristics of two-dimensional membrane airfoils, *Japan Society for Aeronautical and Space Sciences Journal,* Vol. 36, pp. 36–43.

Sussman, M., Smereka, P., and Osher, S. 1994, A level set approach for computing solutions to incompressible two-phase flow, *J. Comp. Phys.*, Vol. 114, pp. 146–159.

Szekely, J. 1979, *Fluid Flow Phenomena in Metals Processing,* Academic Press, New York.

Szekely, J., Evans, J.W. and Brimacombe, J.K. 1988, *The Mathematical and Physical Modeling of Primary Metals Processing Operations,* Wiley, New York.

Tabeling, P., Zocchi, G., and Libchaber, A. 1987, An experimental study of the Saffman-Taylor instability, *J. Fluid Mech.*, Vol. 177, pp. 67–82.

Tao, Y., Sakidja, R. and Kou, S. 1995, Computer simulation and flow visualization of thermocapillary flow in a silicone oil floating zone, *Int. J. Heat Mass Transf.*, Vol. 38, pp. 503–510.

Temam, R. 1978, *Navier–Stokes Equations,* pp. 353–371, North-Holland, Amsterdam, The Netherlands.

Tezduyar, T.E., Behr, M. and Liou, J. 1992a, A new strategy for finite element computations involving moving boundaries and interfaces — the DSD/ST procedure: I. The concept and the preliminary numerical test, *Comput. Meths. Appl. Mech. Engrg.*, Vol. 94, pp. 339–351.

Tezduyar, T.E., Behr, M. and Liou, J. 1992b, A new strategy for finite element computations involving moving boundaries and interfaces — the DSD/ST procedure: II. Computations involving free-surface flows, two-liquid flows, and flows with drifting cylinders, *Comput. Meths. Appl. Mech. Engrg.*, Vol. 94, pp. 353–371.

Thevoz, Ph. , Desboilles, J. L. and Rappaz, M. 1989, Modelling of equiaxed microstructure in casting, *Metall. Trans. A*, Vol. 20A, pp. 311–322.

Thomas, B.G. and Najjar, F.M. 1991, Finite element modelling of turbulent fluid flow and heat transfer in continuous casting, *Appl. Math. Modelling*, Vol. 15, pp. 226–243.

Thomas, P. D. and Lombard, C. K. 1979, Geometric conservation law and its application to flow computations on moving grids, *AIAA J.*, Vol. 17, pp. 1030–1037.

Thompson, J.F., Warsi, Z. U. A. and Mastin, C.W. 1985, *Numerical Grid Generation,* Elsevier, New York.

Thompson, M.E. and Szekely, J. 1989, Density stratification due to counterbuoyant flow along a vertical crystallisation front, *Int. J. Heat Mass Transf.*, Vol. 32, pp. 1021–1036.

Thwaites, B. 1961, Aerodynamic theory of sails, *Proc. Royal Soc. London*, Vol. 261, pp. 402–42.

Tiller, W.A. 1991a, *The Science of Crystallization: Macroscopic Phenomena and Defect Generation,* Cambridge University Press, Cambridge, U.K.

Tiller, W.A. 1991b, *The Science of Crystallization: Microscopic Interfacial Phenomena*, Cambridge University Press, Cambridge, U.K.

Tirmizi, S.H. and Gill, W.N. 1987, Effect of natural convection on growth velocity and morphology of dendritic ice crystals, *J. Crystal Growth*, Vol. 85, pp. 488–502.

Tran–Son–Tay, R., Needham, D., Yeung, A. and Hochmuth, R.M. 1991, Time–dependent recovery of passive neutrophils after large deformation, *Biophys. J.*, Vol. 60, pp. 856–866.

Trivedi, R. and Somboonsuk, K. 1984, Constrained dendritic growth and spacing, *Mater. Sci. Engg.*, Vol. 65, pp. 65–74.

Tseng, A.A., Zou, J., Wang, H.P. and Hoole, S.R.H. 1989, Numerical modeling of macro and micro behaviors of materials in processing: A review, *J. Comp. Phys.*, Vol. 102, pp. 1–17.

Turner, J. S. 1973, *Buoyancy Effects in Fluids*, Cambridge University Press, Cambridge, U.K.

Udaykumar, H.S. 1994, *A mixed Eulerian–Lagrangian approach for the simulation of interfacial phenomena in solidification processing*, Ph.D. dissertation, University of Florida, Gainesville, FL.

Udaykumar, H.S. and Shyy, W. 1994a, Grid supported marker particle scheme for interface tracking, *Num. Heat Transf.*, accepted for publication, (*added in proof:* 1995, Vol. 27, Part B, pp. 127–153).

Udaykumar, H.S. and Shyy, W. 1994b, Simulation of morphological instability during solidification, part I: Conduction and capillarity effects, *Int. J. Heat Mass Transf.*, accepted for publication, (*added in proof:* 1995, Vol. 38, pp. 2057–2073).

Udaykumar, H.S., Shyy, W., and Rao, M.M. 1994, *ELAFINT* – A mixed Eulerian Lagrangian method for fluid flows with complex and moving boundaries, AIAA 94 – 1996, also accepted for publication in *Int. J. Numer. Meths. Fluids*.

Ungar, L.H. and Brown, R.A. 1984, Cellular interface morphologies in directional solidification. The one-sided model, *Phys. Rev. B*, Vol. 29, pp. 1367–1380.

Unverdi, S.O. and Tryggvason, G. 1992, A front-tracking method for viscous, incompressible multi-fluid flows, *J. Comp. Phys.*, Vol. 100, pp. 25–37.

Vanden-Broeck, J.M. 1982, Nonlinear two-dimensional sail theory, *Phys. Fluids*, Vol. 25, pp. 420–423.

Vanden-Broeck, J. M. and Keller, J. B. 1981, Shape of a sail in a flow, *Phys. Fluids*, Vol. 24, pp. 552–553.

Vasilijev, V.N., Dulnev, G.N. and Naumchic, V.D. 1989, The flow of a highly viscous liquid with a free surface, *Glass Tech.*, Vol. 30, pp. 83–90.

Velarde, M.G. 1988, *Physicochemical Hydrodynamics*, NATO ASI Series, Series B, Vol. 174, Plenum Press, New York.

Vere, A.W. 1987, *Crystal Growth*, Plenum Press, New York.

Vicelli, J.A. 1969, A method for including arbitrary external boundaries in the MAC incompressible fluid computing technique, *J. Comp. Phys.*, Vol. 4, pp. 543–551.

Vicelli, J.A. 1971, A computing method for incompressible flows bounded by moving walls, *J. Comp. Phys.*, Vol. 8, pp. 119–143.

Vinokur, M. 1989, Review article: an analysis of finite-difference and finite-volume formulations of conservation laws, *J. Comp. Phys.*, Vol. 81, pp. 1–52.

Viskanta, R. 1990, Mathematical modeling of transport processes during solidification of binary systems, *JSME Int. J.*, Vol. 33, pp. 409–423.

Voelz, K. 1950, Profil und luftriebeines segels, *ZAMM*, Vol. 30, pp. 301–317.

Voller, V.R. and Cross, M. 1981, Accurate solutions of moving boundary problems using the enthalpy method, *Int. J. Heat Mass Transf.*, Vol. 24, pp. 545–556.

Voller, V.R. and Peng, S. 1994, An enthalpy formulation based on an arbitrary deforming mesh for solution of Stefan problem, *Comput. Mech.*, Vol. 14, pp. 492–502.

Voller, V.R. and Prakash, C. 1987, A fixed grid numerical modelling methodology for convection-diffusion mushy region phase change problem, *Int. J. Heat and Mass Transf.*, Vol. 30, pp. 1709–1719.

Wagner, C., Friedrich, R. and Narayanan, R. 1994, Comments on the Numerical Investigation of Rayleigh and Marangoni Convection in a Vertical Cylinder, *Phys. Fluids A*, Vol. 6, pp. 1425–1433.

Wagner, C., Zhao, A.-X., Narayanan, R. and Friedrich, R. 1994, Bilayer Rayleigh-Marangoni convection with and without solidification, accepted by *Proceedings of Royal Society in London (A)*.

Wang, C.Y. and Beckermann, C. 1993, Single vs dual-scale volume averaging for heterogeneous multiphase systems, *Int. J. Multiphase Flow*, Vol. 19, No. 2, pp. 397–407.

Wang, H.P. and Lee, H.S. 1989, Numerical techniques for free and moving boundary problems, in C.L. Tucker (ed.) *Fundamentals of Computer Modeling for Polymer Processing*, Hanser Publishers, New York, pp. 369–401.

Wang, H.P. and McLay, R.T. 1986, Automatic remeshing scheme for modeling hot forming process, *J. Fluids Engrg.*, Vol. 108, pp. 465–469.

Wheeler, A.A., Murray, B.T. and Schaefer, R.J. 1993, Computations of dendrites using a phase field model, *Physica D.*, Vol. 66, pp. 243–262.

Williams, F.A. 1985, *Combustion Theory*, 2nd ed., Benjamin/Cummings, Menlo Park, CA.

Woodruff, D.P. 1973, *The Solid Liquid Interface*, Cambridge University Press, Cambridge, U. K.

Wright, J.A. and Shyy, W. 1992, A pressure-based composite grid method for the incompressible Navier–Stokes equations, *J. Comp. Phys.*, Vol. 107, pp. 225–238.

Xu, J.J. and Davis, S.H. 1983, Liquid bridges with thermocapillarity, *Phys. Fluids*, Vol. 26, pp. 2880–2886.

Yao, L.S. and Prusa, J. 1989, Melting and freezing, in J. P. Hartnett and T. F. Irvine (eds.), *Advances in Heat Transfer*, Vol. 19, pp. 1–95, Academic Press, New York.

Yih, C.-S. 1979, *Fluid Mechanics*, West River Press, Ann Arbor, MI.

Young, G.W. and Chait, A. 1990, Surface tension driven heat, mass and momentum transport in a two–dimensional float–zone, *J. Crystal Growth*, Vol. 106, pp. 455–466.

Youngs, D. L. 1984, Time-dependent multimaterial flow with large fluid distortion, in K. W. Morton and M. J. Baines (eds.), *Numerical Methods for Fluid Dynamics*, Academic Press, New York, pp. 273–285.

Zanio, K. 1978, *Cadmium Telluride: Semiconductors and semimetals,* Vol. 13, R.K. Willardson and A.C. Beer (eds.), Academic Press, New York.

Zebib, A., Homsy, G.M., and Meiburg, E. 1985, High Marangoni number convection in a square cavity, *Phys. Fluids A,* Vol. 28, pp. 3467–3476.

Zerroukat, M. and Chatwin, C.R. 1994, *Computational Moving Boundary Problems,* Wiley, New York.

Zhang, Y. and Alexander, J.I.D. 1992, Surface tension and buoyancy driven flow in a non–isothermal liquid bridge, *Int. J. Numer. Meth. Engrg.,* Vol. 14, pp. 197–215.

Zhang, Y.F., Wang, H.P. and Liu, W.K. 1994, Fast–acting simulation of simultaneous filling and solidification, *Transport Phenomena in Solidification,* C. Beckermann, H.P. Wang, L.A. Bertram, M.S. Sohal, and S.I. Guceri (eds.), HTD–Vol. 284 & AMD–Vol. 182, ASME, New York, pp. 215–226.

Zienkiewicz, O.C. and Taylor, R.L. 1991, *The Finite Element Method,* 4th ed., Vol. 2, McGraw–Hill, New York.

Zweig, A.D. 1991, A thermo–mechanical model for laser ablation, *J. Appl. Phys.,* Vol. 70, pp. 1684–1691.

INDEX

A

adaptive regridding, 200
adaptively moving grid, 122
aerodynamic moment, 70
aeroelastic iteration, 71
aeroelastic parameter, 86
anisotropy, 99, 121

B

backward Euler time-stepping, 112
body-fitted coordinate system, 21, 164, 168,
    172, 186, 187, 197, 198
    continuity equation, 23
    contravariant velocities, 23
    discretized forms, 25
    estimation of metrics, 32
    geometric conservation, 33
    grid velocity, 23
        grid velocity differencing, 26
    jacobian, 23
    metrics, 24
    momentum equations, 24
    moving grid, 23, 24, 35, 70
    nonorthogonal system, 24
boiling, 18
bond number, 39, 42, 48, 49, 55, 57
boundary fitted coordinate system, 108, 109,
    113, 116, 129
    See also body-fitted coordinate system

boundary integral method, 57
boussinesq approximation, 43, 167
breakups/mergers, 202
Bridgman growth, 166, 171, 172
bulk cells, 226
buoyancy induced convection, 38

C

cadmium telluride (CdTe), 166
Cahn-Hillard functional, 14
camber, 86
capillarity, 5, 17
capillary effect, 5
capillary length scale, 122
capillary number, 42, 57
CdTe, 166, 170
circular arc fit, 198
coefficient of solutal expansion, 43
coefficient of thermal expansion, 43
columnar growth, 103
condensation, 18
contact angle, 41, 129
continuous surface force, 10
contravariant velocities. *See* body-fitted
    coordinate system
control volume formulation, moving interface,
    cut cells, interfacial markers, 211
convective fluxes, 216

O

organic melts, 150
  high Prandtl number, 144

P

pattern formation, 103
phase field methods, 14
phase fraction, 152
porous medium, 5
potential flow, 72, 79
Prandtl number, 43, 102, 138, 256
pressure correction equation, 28, 31
pressure-based algorithm, 4, 19
  See also SIMPLE algorithm, ELAFINT
pure material, 103, 138

Q

quadratic polynomial fit, 114
quasi-stationary approximation, 108, 132

R

rapid solidification processing (RTP), 144
Rayleigh number, 43, 47, 101, 138, 256
reference scales, 44
refined ampoule simulation, 178, 182
Reynolds number, 69

S

Saffman-Taylor fingers, 105, 124
sapphire fibre, 128
scale
  heat conduction, 45, 129
  length, 120, 136
    capillary length scale, thermal length scale,
    solutal length scale, convective length scale,
    136
    microscale, 120
  temperature, 109, 120
  time, temperature, 104, 108, 121, 130
  velocity, time, temperature, 44
  morphological length scale, 136
  natural convection scales, 45
  velocity, 121, 129, 141
  velocity scale
    buoyancy velocity scale, 139

interfacial velocity scale, 139
scaling procedure, 44
Schmidt number, 43, 136
second order central difference scheme, 26
second order upwind scheme, 27
sensible heat, 151
SIMPLE algorithm, 26
  coefficients
    momentum coefficients, 26
    source terms, 27, 28
  discretized form, 26
  pressure correction, 28, 30
    coefficients, 28, 29
    source term, 29
  velocity correction, 31, 32
    D'yakonov iteration, 32
single-region formulation, 152
solutal diffusion, 17
solutal expansion coefficient, 43
solutal Grashof number, 43
solute distribution, 101
staggered grid, 216
  See also SIMPLE algorithm
static contact angle, 41
Stefan condition, 107, 108, 109, 136, 147
Stefan number, 109, 112, 113, 115, 118, 121,
    128, 130, 132, 138, 144, 146, 147, 151,
    155, 157, 174, 256
Stefan problem, 156
Stefan-Boltzmann constant, 176
Strouhal number, 69
succinonitrile, 101
supercooling, 103
surface tension, 10, 39, 99, 107, 108, 109, 117,
    120, 122, 123, 124, 125, 126, 127,
    132,133
surface tracking methods, 6

T

T-based method, 155
  See also enthalpy formulation
Taylor-Saffman instability, 18
temperature scale. *See* scale
thermal diffusivity, 54
thermal expansion coefficient, 43
thermal Grashof number, 43
thermal Rayleigh number, 45
thermocapillary convection, 47, 188
tip-splitting, 124
total enthalpy, 156

CPSIA information can be obtained
at www.ICGtesting.com
Printed in the USA
LVOW13s1405140518
577109LV00015B/400/P